THREE GENERATIONS LEFT?

HUMAN ACTIVITY AND THE DESTRUCTION OF THE PLANET

By

Christine Parkinson

ABOUT THE AUTHOR

Dr Christine Parkinson qualified as a Chartered Biologist in 1974, following this with a Master's degree, then a Doctorate in medical research, publishing 50 papers in scientific journals on the way, as well as writing three theses. In 1984, she moved to Birmingham, in a career change, to begin working on inner city regeneration and as a social entrepreneur. She helped to establish three major projects for isolated and vulnerable people, and several spin-off projects from them, in Small Heath, Balsall Heath and Sparkbrook, all deprived inner city areas of Birmingham. The story of these projects is described in her second book, *The Desert will Rejoice* (New Generation Publishing, 2004).

On a sabbatical world trip in 1994 she had noticed clouds of polluting haze hanging over each of the large cities she visited and determined that one day she would act to raise awareness about the factors causing this and global warming. In an *End Piece* to her first book, she described how, from her experience as a scientist and politician, she believed that a whole host of factors were working together to worsen the global climate. This end piece is quoted in the introduction to this, her third book.

Christine has also been politically involved, standing as a parliamentary candidate in 1987 and establishing an organisation, *Christians for Social Justice*, to lobby for the changes to improve social justice in the UK and to encourage Christians to get involved in political issues. She weaves a political dimension into the issue of climate change in this book, as well as drawing on her interest in, and knowledge of, the animal kingdom.

Now retired, in her spare time she works as a trustee for two charities and in fundraising for a project founded by her son to empower African children to become change-makers.

ACKNOWLEDGEMENTS

I am most grateful to Barbara Panvel for support during the writing of this book, by providing links to relevant and supplementary material and for giving a *first read through* of the text of the whole book. Other thoughts and material were supplied by John Nightingale.

Hazel Clawley kindly reviewed chapters 1 and 2 and set me straight on some areas that needed amendment.

To my son, Ben Parkinson, for being prepared to debate some of the more controversial issues, giving me opportunities to clarify my own thoughts.

My uncle, Walter S.F. Bantin, now deceased, provided the funds for the travel I undertook in 1994, during which I first became aware of the clouds of pollution hanging over most of the major cities of the world.

Paper & Print, Shirley provided the graphics for the title page and Fig. 75.

I also acknowledge all those people, whose research, photographs, diagrams and/or cartoons have enabled me to illustrate this book in ways that help with the understanding of it.

CEP 2016

TABLE OF CONTENTS

INTRODUCTION

In 1994 I went on a sabbatical trip around the world, the purpose of which was to meet women engaged in social mission in the cities of this world and to learn from them. It was like a pilgrimage. I met with inspirational women, lived alongside the poor myself and met several visionary people, whose books and projects continue to inspire me. I wrote up the details of this journey in a book and the following (italicised) is a direct quote from my first book[1].

"I met indigenous people and visionary leaders in the Far East, Australia and South Asia. I tasted the colourful diversity and beauty of the world and returned exhilarated. But I also saw much that was not good:

First, and most important, was the pollution suffocating each of the cities I visited, caused by filthy fumes belched from immobile traffic jams, from elderly motor cars, factory chimneys, gigantic airliners and forest fires. An immense, choking, cloud of blue-grey haze hung over each city, destroying the ozone layer, causing global warming, the melting of the ice caps, famine, flooding and changes in climate in all parts of the world – a series of events that may have already gone too far to be reversed.

Second, I witnessed extreme poverty in some parts of the world: mothers with tiny, malnourished babies, begging for scraps, alongside the sick, leprous and disabled, whilst in nearby hotels, the rich gorged themselves in hedonistic displays of self-indulgence. Governments and developed nations seemed oblivious to the suffering of the poor, which I saw as a violation of human dignity, and so little was being done.

Third, I saw that the world had been taken over by greedy merchants. By this, I don't just mean the rip-off merchants to be found in every port of the world; I mainly refer to the people who trade in all kinds of goods for their own benefit, regardless of the effect this has on the stability of the world, its ecosystems, its mineral and animal resources, its local economies and cultural traditions. Examples of this are the following trades: the capture and international trade in rare species, ivory, fur, immature primates etc.; the development of animal foodstuffs from animal carcases (creating cannibalism in ruminant species and diseases like BSE and CJD); the holding of developing countries to ransom by powerful banks, through exploitative usury; a similar use of oil (itself a dangerous pollutant) by oil-producing countries; the development of powerful, polluting and dangerous motor vehicles for their owner's enjoyment; the development of genetically-modified foods for commercial purposes; the unnecessary transport of foods across continents, adding to the pollution and global warming; currency speculation – an international casino in which unscrupulous traders destroy the economies of whole nations; multi-national trading by powerful companies, which destroys local cultures and gains profits by avoiding national controls; the siting of polluting factories close to human populations. The list could go on....

What I see is a blindness of thought, insularity, a lack of responsibility for the wider damage caused by individual actions in all three dimensions, all three inextricably bound together. There is a dogged belief in freedom – but freedom for its own sake, without responsibility, without compassion, runs unchecked. Politicians promote a market economy as if it were a good thing but I saw that it was at the root of the cycle of destruction. Left without controls, it leads to competition and materialism, acquisitiveness spreading like a cancer, greed, the exploitation of one group, nation, or species by another, the concomitant resentment triggering jealousy and wars, with the end result being the ultimate destruction of our beautiful world by selfish people. Rather than assessing needs to develop standards, values and strategies, the vagaries of the market determine priorities and direction, so that the rich benefit at the expense of the poor, as well as at the expense of the planet…..

Tinkering with these issues at G8 summits is not enough. Resolving conflicts of interest through threats and wars, and the offering of paltry aid to poor countries are also totally inadequate. To reverse current trends, and to prevent the destruction of the world, there is an <u>urgent</u> need for co-operation between nations, in which the commonality of the human condition is stressed, rather than its diversity. Then, mankind might find a way to tackle global warming, to alleviate extreme poverty and to frustrate the exploitation of the merchants. One vehicle for this might be through a reformed and inclusive United Nations, commissioned to act with robust determination to save the planet, its peoples and all its beautiful creatures."

The things that I saw and the people that I met during those months of travel continue to influence and encourage me. On my return to the UK, inspired by what I had seen, I went on to set up two more social projects for marginalised people living in the inner city of Birmingham. The setting up of these projects is what has delayed me in the writing of this book. It has taken me 22 years to actually put pen to paper about the issue of climate change.

At about the same time as I was writing my first book, I also received a number of letters from a friend and neighbour, Barbara Panvel, who divided her time between living in Mumbai, India, and in the UK. One day she was meditating whilst in India and felt compelled to write to me about a new initiative that she was being led to develop. She sent me a *New Era Statement* (see Table 1) for my comment. At that time, she hoped to establish a coalition of like-minded people who would identify with the statement. She later set up a *New Era* Network website (www.neweranetwork.info) and, initially involved several people from her wide circle of contacts in writing reports on "Counting the Costs" (of our present system); these reports are made available on the website. The first of these, written in 2005 by Jeremy Seabrook, Mark Tully and Molly Scott Cato (now a Green MEP)[2]. I will draw on this material in the writing of this book, as it provides statistics which give weight to the theory that I will attempt to expound.

More than ten years have passed since Barbara set up the website and it is still active, containing a wealth of positive information, added to daily, about initiatives that may bring about the changes to our society that we seek.

As we continued our correspondence between Mumbai and Birmingham, we came to an understanding that, rather than listing all the negatives about our present dysfunctional society, we needed to define what a civilised society might look like. So, we put together a list, which each of us has used in our separate ways since then. Barbara constructed several websites associated with the list and adds material to them regularly. The websites are within:

https://civilisation3000.wordpress.com/ and https://civilisation3000.wordpress.com/five-civilisation-3000-sites-relate-to/

and include the following:

Civilisation 3000: civilised attitudes to defence

Economic & political affairs: plans proposed & measures taken to improve the political and economic life of the country

Antidote to doom and gloom: environment and beneficial innovation

Food production – a vital public service: the provision of food and water in this country and worldwide

United Nations: not them but us: human rights

The sites are visited regularly by hundreds of people from across the globe. It is a valuable piece of work and I have drawn on articles posted on the websites in the writing of this book.

I expanded the list of the characteristics of a civilised society to a total of 17 factors and published it at the end of my first book (shown here in Table 2). Some might think this list to be somewhat Utopian – and it probably is, as there is no doubt that no society exists today, anywhere in the world, which could tick all 17 boxes. Maybe there are some which could tick none of them. But we need a model to aspire to and this list is as good as any other.

In 1979, James Lovelock, an eminent British scientist, published a theory, which he named "Gaia"[3]. It is an inspirational and classic work, which sees the evolution of life and the evolution of the Earth as a single, tightly-coupled process from which the self-regulation of the environment emerges. It is now available in paperback. As a scientist myself (life sciences), I find his theory fascinating and compelling, though it has been criticized by some scientists. As a Christian who strongly believes in good stewardship of the earth, I also find it exciting. The main thrust of Lovelock's theory is the inter-connection of all the ecosystems on earth and in the atmosphere. If one thing goes out of kilter, such as too much carbon dioxide in the atmosphere, then the whole of life on earth will be affected. And we are living in such a time, with global warming caused by an excess of carbon dioxide, leading to climate change and catastrophic weather events and the loss of indigenous species, with the eventual destruction of the planet, and the life on it, if nothing is done to stop the process by re-balancing the systems and cycles and reducing carbon emissions.

It is my belief that the social, demographic, industrial, trading and economic systems on this earth have a similar inter-connectedness, with each other and also with environmental ecosystems. The purpose of this book is to try to demonstrate this inter-connectedness. I am a scientist and a social entrepreneur, not an economist, but I will draw on other people's theories in those areas where I am less qualified.

6

We have all seen experimental set-ups with dominos – thousands of them set up in a pattern across a room – then one domino is pushed over, to start a chain reaction in which they all eventually tumble over. I believe it is the same with the inter-connected systems of this world – if one gets out of balance, then all the others will eventually follow, tumbling down into chaos, like a house of cards.

All of the inter-connected factors will be dealt with separately, in their own chapters but I hope to show in this book how they each overlap, connect and influence all the others. There are certain books, which I have drawn on particularly and I would recommend them to anybody who wishes to study this subject further. They include:

Winin Pereira and Jeremy Seabrook (1996) *Asking the Earth - the spread of unsustainable development*[4];
Richard Douthwaite (1999) *The Growth Illusion*[5;]
Paul Rogers (2012) *ORG Special Briefing*[6;]
George Monbiot: www.monbiot.com and his various articles in the Guardian[7;]

James Lovelock (2009) *Gaia*[3].

And, in addition to these, I must add Naomi Klein's book, *"This Changes Everything"*[7]. Her book came to my attention when I had written about half of my book and it is rapidly moving to the position of being a classic work on climate change. I refer to some of her ideas in the later chapters of my book.

A learned colleague of mine, who was also an editor, once advised me that, when writing up my scientific research for publication, I should aim to explain things so simply that an elderly aunt might be able understand them. Whilst his comments were both ageist and sexist, I have understood the spirit of his comment and I still try to take his advice when I write. Much of this book is therefore written in a fairly simple style, in the hope that non-scientists will be able to understand it. I believe this is important because, whilst there are many people in the world who read, and link into the New Era Network and Civilisation websites, there are many more others who seem unaware of the dangers that we face on this planet, if we continue with our present way of life. Maybe these others have been influenced by the right wing press into thinking that what the scientists are saying is not true or is overly exaggerated. I would like this book to be read, and understood, by many people, for it is only when the masses come together to lobby for change, that change will occur. The political powers, both nationally and globally, have too much vested interest in maintaining current systems and economies, to bring about the changes that are urgently needed. Their links with the business world also affect their ability to see the situation holistically and objectively.

But I apologise in advance to those people who find this book too simple or who believe that I am telling them what they already know. If you are one of these, please bear with me as there may be something new for you here among the pages – or material that you can use in your own awareness-raising and/or lobbying for change.

TABLE 1

NEW ERA STATEMENT
(from 1999)

"Actively, or through our inactivity, we have fashioned a society whose members are fed on polluted food, breathe toxic air, have varying degrees of ill health or disease due to damaged immune systems, and are distressed by rising levels of violence – nationally and internationally.

Having historically deprived our young people of the ability to be independent by taking their share of the commons – land, skills and the means of production – all we can currently offer them is a poisoned chalice, containing no hope of a better future and the abhorrent legacy of nuclear waste.

To enter a new era by effecting beneficial change, we think it important that people rescind their acceptance of the party political system and of the current economic mythology which has brought about our present social, psychological environmental stresses and distresses.

The decision-making, which has led us into our present situation, has been conducted within the framework of a party political system, which attracts people, greedy for power and/or money, to come forward and govern. Most able, honest and altruistic people will not contemplate taking part in it as presently constituted.

Conventional classic economic theory and practice ignores the basic rights of the majority in order to make huge profits for a minority – similarly greedy for power and money.

People who are altruistic, demonstrably honest (in financial and personal dealings), aware that every one of our institutions is failing and are seeking wholesome alternatives, need to come together…..

A New Era Movement would grow organically….. Its emphasis would be to promote and celebrate ethically sound, socially just, environmentally sustainable developments in every sphere."

TABLE 2

NEW ERA: CIVILISATION 21st CENTURY

In a civilised society people should have:

1. Pure drinking water;

2. Clean air;

3. Sufficient food to keep them from starvation, that is uncontaminated by pesticides, genetic modification, food-chain-infringement or other pollution;

4. Adequate shelter from the elements;

5. Access to good, affordable medical care;

6. The opportunity of high quality education throughout their lives;

7. Opportunities to participate in self-fulfilling work;

8. An expectation that they will be cared for in a thoughtful, un-violating way, should they become orphaned, injured, disabled or disadvantaged in any way;

9. A choice of sustainable livelihoods;

10. Freedom of (courteous) speech;

11. Effective participation in decision-making;

12. An agreed code or legal system, effectively and democratically implemented;

13. Freedom of association (political, religious etc.);

14. An ability to move within society without fear of violence, abuse, derogation, marginalisation or exploitation;

15. Respect for the needs, dignity and rights of other peoples and species inhabiting the globe;

16. Honesty and transparency in both financial and personal relationships and dealings;

17. Opportunities to resolve problems through dialogue and non-violent action – at local, regional, national and international level.

CHAPTER 1
Our beautiful world in harmony

One October, when I was about 6 years old, my mother took me out for a treat. My older siblings were involved in other things and this was a rare opportunity for me to have my mother's undivided attention. We walked to a local park, Scotch Common, which had a variety of trees, beginning to show their autumn colours: coppers, browns, golds, ochres and reds. We identified some of the trees as horse chestnut, oak and sycamore and then searched beneath them to collect their seeds: shiny brown conkers with a varnish-like sheen, green and brown acorns, some separated from their craggy cups, and the winged sycamore paired seeds, which would spiral slowly down to the ground if you threw them into the air. Mum suggested I take a selection to school to put on the nature table.

I don't know why this incident sticks in my mind but I believe that it may have been the beginning of a growing love of nature in me, which is still a significant part of my identity. Though I am now 72 years old, each autumn I still collect conkers and acorns and sycamore seeds for my own nature table at home. I don't know how much longer I will be able to do this, as the seasons are changing so much. Already the conker crop this year seemed smaller and autumn was extended with a mild spell, with golden leaves on the trees until well into November, and winter still not started by Christmas. Are we in danger of losing some of these great trees and their fruit and their annual cycles related to the seasons? Why is it that we have summer flowers still in blossom in December and reports that in some parts of the UK, the spring flowers (daffodils etc.) are already in blossom in December?

Yes, I love nature but my love of animals far surpasses that of the plant kingdom. We share this world with some wonderful creatures: the large wild carnivores and herbivores of Africa and Asia; the strange marsupials of Australasia; the prairie animals; the domesticated pets who share our homes with us; the birds who visit our gardens and who migrate across great oceans every year; the creatures and fish of the seas; the inhabitants of the polar ice caps and the smaller secretive wild mammals who live in burrows.

I believe that I am not the only person in this world who loves nature in this way and who respects and enjoys the splendour of our world. We live on a magnificent planet and share it with some spectacular creatures.

I am writing this book because I believe that we are in danger of losing it all. And the magnitude of this loss is greater, and the need for action more urgent, than many believe.

How everything fits together in harmony

It has been known for more than 50 years, and certainly since I was at school during the 50s and 60s, that the process of photosynthesis in plants is closely linked to the process of respiration in animals. Indeed, one could almost describe the relationship between plants and animals as symbiotic, one being dependent upon the other to maintain its life. The plant life on the planet absorbs carbon dioxide from the air and water from the soil and, through chemical reactions, changes them into glucose and oxygen. The oxygen is released into the air and breathed in by the animal life (including ourselves). In animals, oxygen is inhaled and carbon dioxide is released through the process of respiration. Thus, plants provide oxygen for animals to breathe and animals exhale carbon dioxide, which is used by plants in the process of photosynthesis.

This photosynthetic cycle has been analysed and shown to be a series of chemical reactions, all initially triggered by light energy from the sun. Chloroplasts in plants (in the green chlorophyll) trap the sunlight, which provides the energy for the photosynthetic cycle (Fig.1)

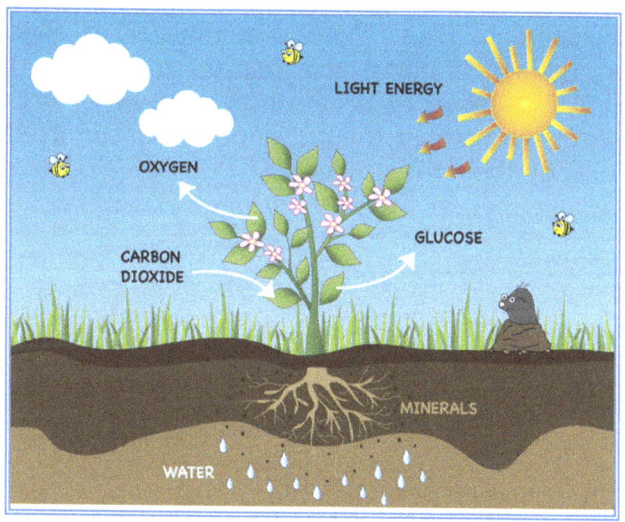

Fig 1. The relationship between photosynthesis in plants and respiration in animals
From: http://www.motherearthnews.com/nature-and-environment/nature/how-photosynthesis-works-zw0z1406zwea.aspx with permission

This process happens throughout nature, from the very smallest algae and plankton to the giant trees in our forests and from the smallest amoebae and zooplankton in water to the largest of our land and sea mammals (elephants and whales) - an interchange of gases and chemical products between plants and animals which is important to sustain life.

But the photosynthetic and respiratory cycles do not stand alone. They are inter-linked with other kinds of cycles, the chemical processes of which have been carefully studied by scientists. For example, plants store another product of photosynthesis (glucose or starch) and this is consumed by herbivorous and omnivorous animals and provides them with the energy they need for growth and development. Thus, there is a transfer of energy from the sun to plants and then on to animals, this energy is needed to sustain life. And none of this could begin without the presence of the sun itself – at exactly the right strength.

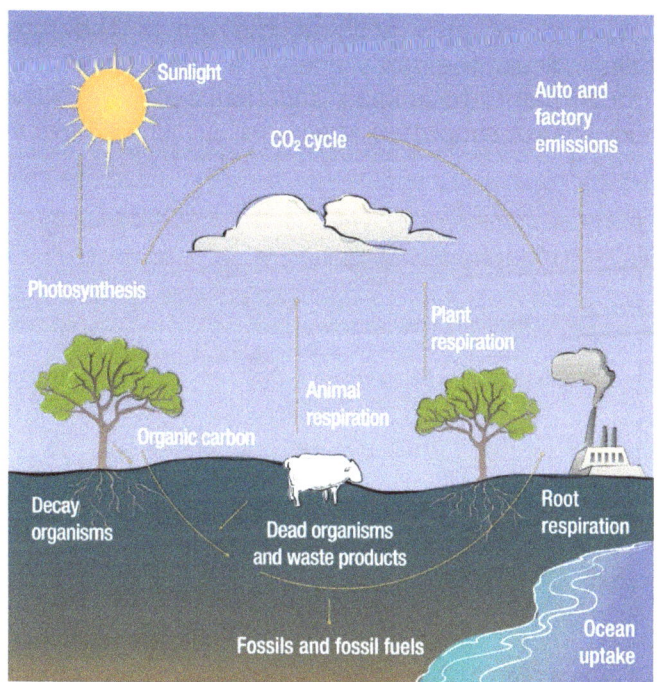

Fig 2 THE CARBON CYCLE
Illustration by LizzardbrandInc, with permission from
UCAR

There are other cycles in nature too: the nitrogen cycle, the Krebs cycle (to process and release energy) and the carbon cycle (Fig 2), which is closely linked to the respiratory cycle of animals. The carbon cycle involves the decomposition of dead and decaying matter into fossil fuels (see later for the significance of this).

Following the discovery of interactive cycles in nature, it was not long before the whole concept of food chains was proposed, with the lowest forms of life being consumed by the next species up the food chain, from herbivores (plant eaters) to omnivores (plant and meat eaters), with the carnivores (big cats, birds of prey etc.) at the top of the food chain. Thus, the sun's energy is transferred first through plants to animals and then up through the food chain, simplified diagrams of which are in Figure 3.

Fig 3. Simple Food chains
From: www.k8schoollessons.com/food-chains-and-food-webs/ with permission

The diagram in Fig.3 shows simplified food chains but, in fact, things are rarely as simple as this and the concept of a "food web" is much closer to reality. Figure 4 shows a woodland food web, which can be seen to be much more complex than a simple chain, with various species being inter-dependent.

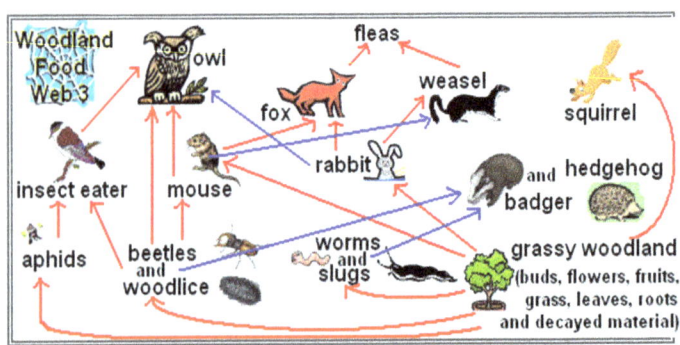

Figure 4: A Woodland Food Web from
www.docbrown.info with permission

A recent programme on BBC TV, "Secrets of our Living Planet", also available on DVD[9], gave examples of some fascinating food webs throughout the world, from tropical rain forests, to savannahs and in the oceans, and demonstrated that if one member of the web disappeared, then others wouldn't survive. The most compelling example of this was the brazil nut tree, which relied on a small rodent, the agouti (Fig. 5), to crack and disperse its seeds, as well as an orchid, which grew on its trunk and attracted a particular species of bee, to pollinate both tree and orchid, the male bee pollinating one and the female bee pollinating the other, with the bees reliant on the nectar in the flowers for their survival.

Fig 5 - the Brazilian agouti (from
www.hidephotography.com with permission)

So we can see from this that, not only is there an interaction and inter-dependency between plants and animals, but that inter-dependency continues throughout the animal kingdom, in a complex web. Thus, if one species disappears, or becomes extinct, this may also affect other species, which are dependent on it as a food source or pollinator. This whole interaction between members of the plant and animal species is called an ecosystem.

I feel that the interaction of all the cycles and ecosystems is close to being miraculous. Our world has been regulated in an astounding way. It is as if everything on this planet has been put in place in ecosystems, or has evolved, to work harmoniously, so that all life on this planet remains in balance, in a wonderful connectedness and interdependency that maintains life.

I love to wander through parts of our green land, with rolling hills and tranquil forests, just taking in the beauty of it. I also love to visit beaches to hear the sea and breathe in the clean, salty ocean air. It is not surprising therefore that I have been excited by the hypothesis proposed by the scientist, James Lovelock, in 1979[10], which states that the earth itself is a self-regulating body; that the earth is like one big organism with the ability to regulate critical systems to meet its own needs and to sustain life. It is called the Gaia Hypothesis.

Gaia Hypothesis

An idea proposed by James Lovelock (1979)

All living things on earth (biosphere) function as <u>one</u> SUPER organism that changes its environment to create conditions that best meet its needs, with the ability to self-regulate critical systems needed to sustain life.

Fig.6 Gaia Hypothesis

Whilst I would not go so far as Lovelock in concluding that the earth acts like one super-organism, I do believe that there is a synergy and harmony about the interconnection of all the ecosystems operating across the globe.

The regulatory mechanisms which have been keeping all life in balance and harmony for thousands of years are now being undermined and put out of harmony by the hand of man. Let's have a look at what we have been doing to place all this at risk and what we need to do to make things right again.

Our beautiful world no longer in harmony

Fossil fuels, produced as part of the carbon cycle, have been used by humans for centuries, but especially since the industrial revolution, to produce other forms of energy for humans to heat their homes, run their vehicles, power up vast factories and to develop more and more complex gadgets and life-enhancing commodities. The downside of this practice is, of course, that carbon dioxide and other toxic gases are released into the atmosphere as a by-product of their use, resulting in global warming.

Global warming is the rise in average global surface temperature caused primarily by the build-up of human-produced greenhouse gases, mostly carbon dioxide and methane, which trap heat in the lower levels of the atmosphere.

At the beginning of the industrial revolution it was not realised that the plant life on earth could not cope with absorbing all the extra carbon dioxide being released into the atmosphere from manufacturing and the problem was made worse by the felling of many of the great trees in the mighty rainforests of the earth, in order to clear land for agriculture and to sell the wood. Figure 7 shows the dramatic increase in fossil fuel emissions since 1870. This is comprised mainly of carbon dioxide.

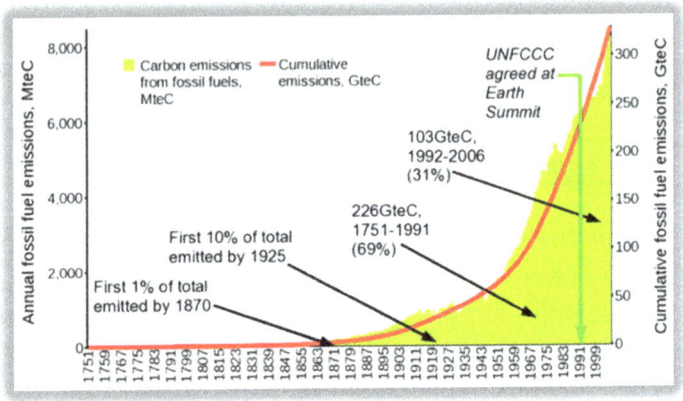

Fig.7 Fossil fuel emissions since 1751
From:
http://www.fraw.org.uk/mei/ecolonomics/00/ecolonomics-20091013.shtml
GteC refers to Giga-tonnes of carbon

Human activity has been bringing all the ecosystems on the planet into an imbalance and a resulting effect of this has been the loss of numerous species, as well as changes to the climate and global temperatures.

Another way in which plant life and animal life (insects and birds) have interactive cycles is the way in which bees depend on flowers for nectar and, in visiting plants to feed on nectar, they inadvertently brush against the pollen in the flower stamens. They then carry this pollen on their bodies to other flowers and become the means by which pollination occurs in plants (part of the reproductive cycle of plants). Recently, vast decreases in the numbers of bees have been noticed and this is thought to be caused by the use of pesticides on plants. If the bees were to disappear altogether, pollination might not occur and this could reduce some of the food sources available to us. Vegetables and fruit known to be pollinated by bees are okra, kiwifruit, onion, celery, cashew nuts, strawberries, papaya, custard apples, turnips, beet, Brazil nuts, carrots, broccoli, cauliflower, Brussels sprouts, cabbage, water melon, coconut, tangerine, cucumber, quince, fig, apple, walnuts, mangos, avocados, peach, nectarine, pear, raspberry, blackberry, elderberry, cocoa, passion fruit and many others.

Thus, the loss of bees might result in the loss of most of the vegetables and fruits that the human race, and other species, rely upon for their food.

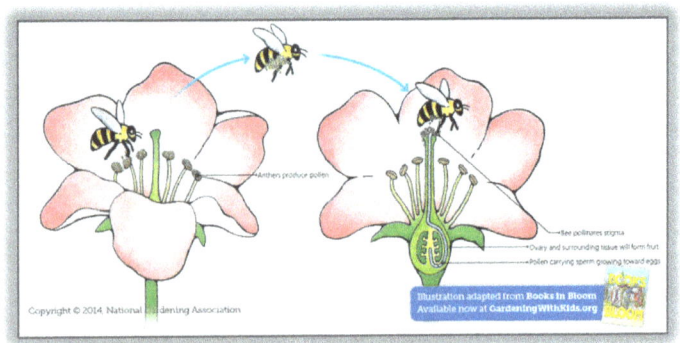

Fig 8: Bees in the process of pollinating flowers
From: http://www.kidsgardening.org/node/99559

There have been vast changes in the way that farmers have carried out their agricultural activities in recent years; they have copied some processes from the manufacturing industry to become more "productive", using intensive farming methods, removing hedgerows and maximising the use of their fields. **Over this same period, certain species of birds have been disappearing because the insects in hedgerows that they feed on are no longer there, or have been killed off with pesticides.**

Wikipedia lists 190 species of birds which have become extinct since 1500 and a further 321 are currently endangered, including the cuckoo and several of our garden species.

A recent report from American scientists, Ceballos and colleagues[11], suggests that human activity has already triggered the beginnings of another mass extinction, thereby threatening our own future. According to this group, there have been five mass extinctions in the earth's past (the extinction of dinosaurs being the most well-known) and that this latest threat to the planet would be its sixth mass extinction. They state that, in the last century, vertebrates (animals with backbones) have been disappearing at a rate 114 times greater than would normally be expected, without the destructive activity of humans. They pointed out that, since 1900, over 400 more vertebrates than expected had vanished; this included 69 mammals, 80 birds, 24 reptiles, 146 amphibians and 158 fish species. **They warn that species loss will have a significant effect on human populations in as little as three generations.** The researchers concluded that this destruction of species is accelerating and initiating a mass extinction episode unparalleled for 65 million years.

This report has triggered significant discussion within the scientific community, and some have ventured to include humans (also vertebrates) as part of this extinction. They are confident that bees will definitely be extinct by then and perhaps many of the large carnivores, such as lions. Whether humans also become extinct depends, one supposes, on whether those creatures and plants which we rely on for food, have disappeared in this mass extinction. It is estimated that 2,000 sheep and 100 cattle were drowned in the recent floods engulfing the north of England, so the loss of our food sources due to climate change is a possibility. So, with bees gone and the vegetables that they pollinate and the loss of some of our meat sources, things look bleak for humans in the future too. A number of organisations are predicting crop failures due to climate change by 2030, particularly in the poorer countries in Asia and Africa.

There are also concerns about the effects of climate change on human health[12]. This 43-page significant publication by Antony Costello and others gives evidence of grave concern to human health.

Anthony Costello, director of the UCL Institute for Global Health said: *"Our analysis clearly shows that by tackling climate change, we can also benefit health — and tackling climate change in fact represents one of the greatest opportunities to benefit human health for generations to come"*. And Hugh Montgomery who co-chaired the Commission said, *"Climate change is a medical emergency. It thus demands an emergency response, using the technologies available right now"*.

Also, in its 2010 report *"A Human Health Perspective on Climate Change"*[13] the National Institute on Environmental Health Sciences gives a list of the health consequences of increased greenhouse gases and climate change. The list includes about twelve major health risks. The human population would therefore seem to be as much at risk as the creatures with whom we share this planet.

And yet, humans don't seem to be able to stop tinkering with the natural order of things in the ecosystems of the world. One vivid example comes from Australia where, in 1935, a toad from South America was introduced to Queensland, with the aim of using it to consume cane beetles, which were damaging sugar cane crops. This toad did not eat the beetle and instead multiplied in huge numbers, because it had no natural predators, so that the cane toad is now a national pest. It is also poisonous to other species and is now being blamed for a massive reduction in the number of dwarf crocodiles in Australia.

Fig. 9: Cane Toad

To go back to farming practices: Fields are no longer left to lie fallow and so do not have a chance to replenish the nutrients found in soil that are essential to plant life, so that they become less productive. However, some farmers are now introducing permaculture, with good results and organic farming is also on the increase.

During the 1990's the condition of "mad cow disease" (BSE – bovine spongiform encephalopathy) appeared in the UK and it was eventually discovered that foodstuffs fed to cattle at that time had been processed from animal sources and so cows, who are herbivores, were being fed foodstuffs which turned them into not only carnivores but also cannibals. This violation of the natural food chains had far reaching consequences, as it would appear that it could be passed on to humans who consumed meat from cattle with BSE, the human form of the disease being named CJD (Creutzfeldt-Jakob Disease).

Another example of human activity which had devastating effects on the life of the planet.

In a recent *"Springwatch"* programme on BBC TV, we were made aware of another dangerous practice: the production of exfoliation products for washing our faces; these soap-based products contain tiny particles of plastic (which do the exfoliation); these are washed down sinks and eventually get down via rivers into the sea. They are absorbed by micro plankton, which are subsequently eaten by fish – and thus find their way into the food chain, if they do not kill the fish off first.

So here we have several kinds of human activity that are interfering with the natural cycles and transfer of chemicals and energy through the plant and animal kingdoms, as well as through the food chains:

- the whole industrialisation process, which releases excessive carbon dioxide and other toxic pollutants into the air;
- the use of pesticides to enhance agricultural production, which has killed off bees and other insects and also birds;
- intensive farming methods which have eliminated hedgerows and thus the bird species which rely on them for nests and food;
- the feeding of processed animal products to herbivores;
- the expansion in the use of exfoliants, which get into rivers and seas and work their way up through the food chain;

- the introduction of non-native species into other countries;
- deforestation and land clearance.

And these have not been the only human activities to do this. Humans also **exploit** the animal kingdom, sometimes in very cruel ways, in order to make money for themselves and this has also put some species at risk of extinction. This exploitation includes killing elephants for ivory, rhinos for their horns, sharks for their fins, bears for their bile, pangolins and forest mammals for their meat and capturing baby monkeys and other primates, some from rare species, to sell in markets. Some species, such as the tiger, are currently threatened because of **habitat loss or fragmentation**. Forests where the tiger lives are cleared for agricultural activity, such as growing palm oil. Many other species are also in danger because of habitat loss (orangutan, elephant, rhino, polar bear etc.). As I write, we hear about a huge fire in the country of Indonesia, originally started to clear forest for the planting of palm oil crops, but now burning out of control, leaving a smoky haze over a wide area. Indonesia is the only habitat for the endangered orangutan, as well as the rare Bornean white-bearded gibbon, sun bears and pangolins.

Global warming has led to the melting of the ice caps and a subsequent rise of sea levels, so that some island nations are at risk of disappearing into the sea. Scientists have predicted that global average surface temperatures are likely to rise by 3-4° within the lifespan of today's teenagers, though there are efforts to keep it down to below 1.5°. The BBC recently reported that, as 2015 has been a particularly hot year, the average global temperature is likely to increase above 1° for the first time[14.] In a later chapter I will discuss the efforts being made at UN level to keep the temperature rise below 1.5°.

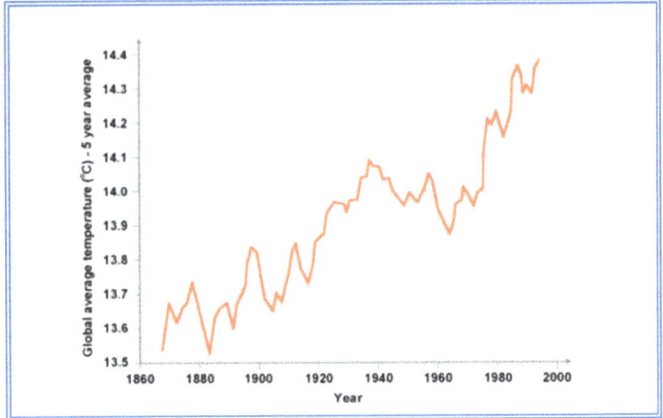

Fig. 10 – increases in global average temperature since 1860
From: www.bbc.co.uk (GCSE Bitesize)

Recent reports, described in the Guardian have demonstrated that global temperatures in February and March 2016 have been the hottest since records began[15] and 2015 was the hottest year on record before that.

Also affected has been the climate, with more frequent catastrophic events, such as tornados and cyclones, mud slides, flooding, droughts, desertification etc. With the melting of the polar ice caps, the ecological balance of species living in these areas has also been disturbed, the most well-known being the polar bear, which can no longer rely on its main food resource, the seal.

FIG 11: A starving and emaciated female polar bear on a small block of ice
Photograph by Kerstin Langenberger with permission

Already, in several parts of the world there has been a rise in sea level, affecting especially coastal areas and island nations (Maldives, Marshall Islands, Philippines, Tuvalu, Solomon Islands). Over the past century, the world's oceans have risen 4-8 inches. It is reported that several rocket launch areas and space stations in the US will have to be moved inland, because of the risk of flooding. Scientific models have suggested that sea levels will rise by 20 centimetres by 2050 (that's another 8 inches), or triple that if the ice sheets in Greenland and Antarctica continue to melt. The acidity of the sea has also increased by 30%, due to it absorbing carbon dioxide from the atmosphere to form carbonic acid and this puts some marine creatures and coral at risk. Coral reefs are particularly in danger, especially the iconic Great Barrier Reef, just off Australia. Australia's recent surge in industrialisation projects (mega-mines, dredging and railway projects) has put the reef in danger with rapid destruction of the coral. We are told that 50% of coral has been lost since the 1980s, due to the warming of the sea.

Fig. 12: Bleaching of the coral in the Great Barrier Reef
From: https://fightforthereef.org.au

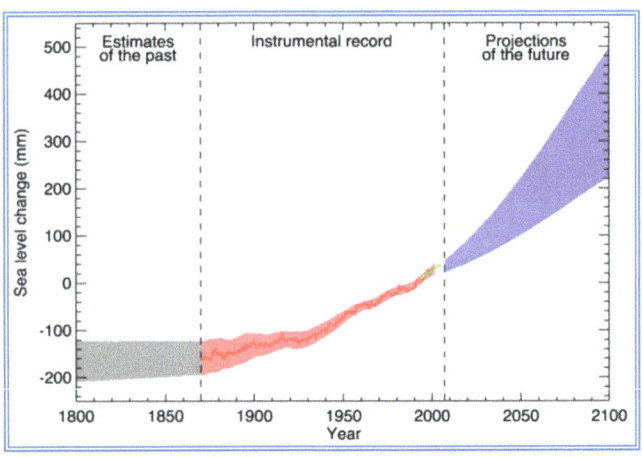

Fig. 13 – the increases in sea level over the last century
Source: US EPA Climate Change Website

Most worrying are the vast permafrost regions of the world (Siberia, Canada, Alaska), where the earth remains below freezing point, even during the summer. If the temperature of these areas increases, then large quantities of methane will be released into the atmosphere, adding to the problems of global warming that we already have.

At this time, there are campaigning groups trying to stop companies drilling for oil in the Arctic ocean, where the sea ice is already melting at a rapid rate. Recent studies in Greenland have shown that 2014 was the warmest year on record going back 135 years, with evidence that the glaciers are shrinking and the ice is thinning – and 2015 has been shown to be even warmer. A recent report from the Californian Institute of Technology states that one of the biggest glaciers in Greenland, Zachariae Isstrom, which holds enough ice to raise the sea level by 18 inches, has broken loose from a stable position and is melting at both ends, with ice crumbling into the North Atlantic Ocean. Greenland is the second largest ice body in the world and already contributes to about 40% of the current sea level rise. Since 1992, 65 million tons of Antarctic ice has melted.

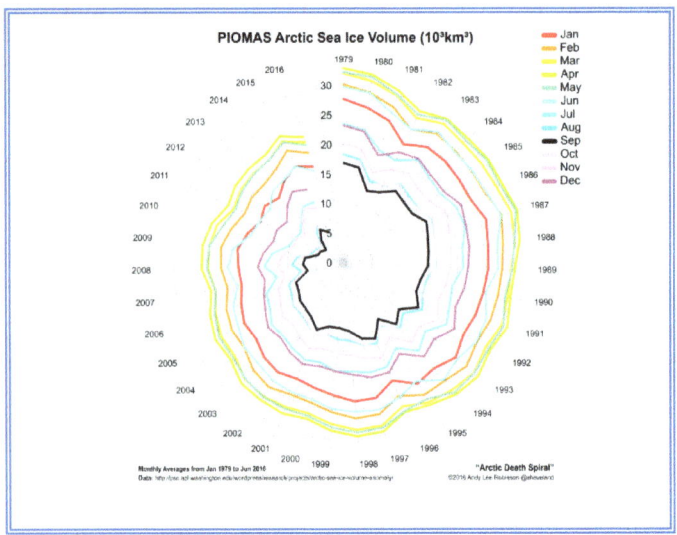

Fig. 14: The Shrinking of Arctic Sea ice between 1979 and 2016 © Andy Lee Haviland

Some have made calculations about what would happen to the world if all the ice caps were to melt and it is quite clear that, not only island nations, but also whole countries and some major cities would be swallowed up by the sea. The maps in Fig.15 (from National Geographic) show what would happen to the world in that circumstance. Such a circumstance would remove much of the UK, especially the eastern areas and around the wash and the Thames, the whole of the Netherlands, almost all of Denmark, part of northern Germany, much of Turkey and the Baltic regions. Venice would disappear into the Adriatic Sea and the Caspian and Black Seas would become much larger. Worldwide, we would lose Bangladesh, Singapore, some of the Philippine Islands, much of Sumatra and Papua New Guinea, the whole of Florida, several Caribbean islands, Tuvalu and much of China. Huge inland seas would develop in Australia, around the

Amazon and Paraguay River basins and delta areas would also be inundated (Mekong, Nile, Ganges), leading to the submergence of Cairo and Alexandria. Due to differences in ocean currents, the sea level increase would be higher in some areas than others (e.g. the eastern seaboard of the USA). Africa's coastline would not be as affected as that of some other continents but, due to temperature rises, some parts would become so hot that they would be uninhabitable.

All of the changes described above have not gone unnoticed and there have been numerous campaigns and demonstrations to prevent some of the human activities which are endangering our planet, some more successful than others. For example, in the Netherlands, one of the countries most at risk of rising sea levels, the Hague District Court recently ordered the Dutch government to reduce the country's greenhouse gas emissions by at least 25% by 2020. This arose following a complaint by an activist group. The Netherlands is particularly vulnerable to the effects of global warming, with much of its land lying below sea level. The island nations are also at risk of being swallowed up by the sea but this is as a result of greenhouse gas emissions of other countries, rather than their own. And the Philippines were recently devastated by Cyclone Yolande.

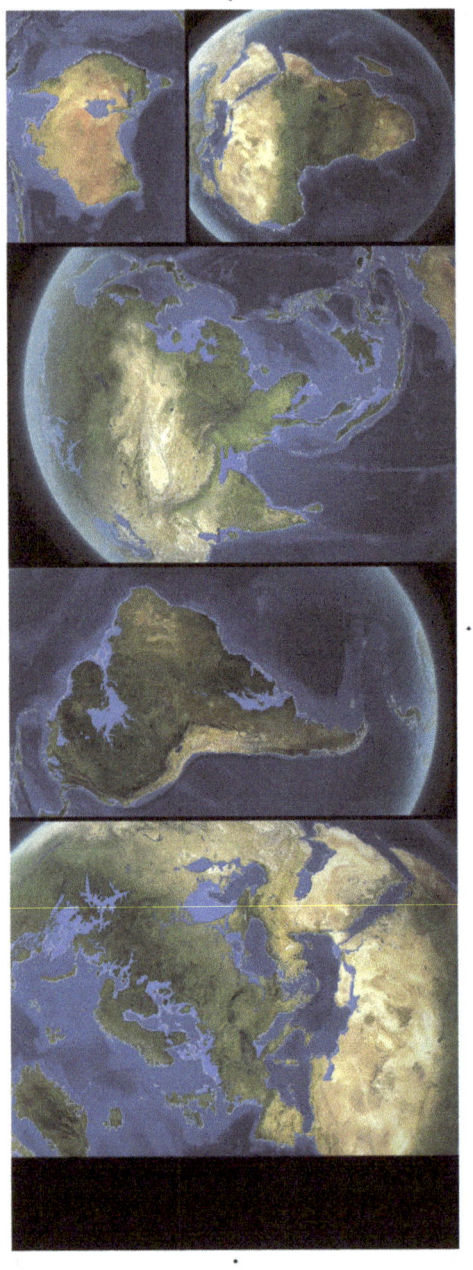

Fig 15: Pictures showing the new coastlines if all the ice caps were to melt - the outer lines show coastlines as they are at present. Source: National Geographic Creative.

Fig 15a. New Coastlines of Europe if all the Icecaps were to melt. From: National Geographic Creative, with permission

The United Nations has been taking action, ever since the Rio Summit in 1992 (to be described in a later chapter) but it is not enough, as carbon emissions continue to rise. As I started to write this book, the latest summit (CPO21 in Paris) had not yet taken place but, by the time it was finished, an agreement had been reached, which will be discussed in Chapter 8.

Some of the toxic chemicals released from human manufacturing activity, such as nitrous oxide and bromine and chlorine compounds (CFCs), have the effect of depleting the ozone layer, which exists in the earth's atmosphere. The purpose of the ozone layer is to absorb ultra-violet rays from the sun. Ozone levels in the stratosphere have reduced by 4% since 1970 and there is an ozone hole over the Antarctic Circle – again more evidence that human activity is affecting the stability of the planet.

So many factors have been interacting to create the global situation in which we are at the moment and this book attempts to show how they interrelate. Each chapter in this book will look at a different factor, which has put the planet and its species at risk and will show, I hope, that each of these has an inter-connectedness. We therefore need to tackle every factor, not just one in isolation. Scientists have said that we have only **three generations** to do this before things have gone too far. If the graphs shown in Figures 7, 10 and 13 continue at this rate, then we probably have even less time than three generations to reverse the changes.

A short piece of film has recently been circulated on the internet, which summarizes all of these risk factors, and is especially targeted at those who, like me, love and cherish the natural world[16].

Scientists have predicted that, in three generations time, there will be a mass extinction of many of the animal species inhabiting this planet. It is not clear whether this extinction will include humans but many of the animal, insect and bird species that we have grown to love will have gone by then. I think the risk is there for human populations as well, so I have used this "3 generations" factor as the title of this book and in most of the assessments and discussions which follow. Let's hope that this never happens but using *3 generations* as a rule of thumb will hopefully concentrate the minds of those who are in positions in which they can make the changes needed to ensure that this never becomes a reality.

SUMMARY OF INTERCONNECTIONS DESCRIBED IN THIS CHAPTER

CONNECTION	AS SHOWN BY
1.Dependent relationship between plants and animals	1. Cycles of the natural world: Photosynthesis; respiration; carbon, nitrogen and Krebs cycles; food chains and food webs.
2. Climate change and global warming	2. Human activity causing increased carbon emissions into the atmosphere and the sea and a greenhouse effect;

	Human activity causing the release of other pollutants (CFCs etc.) into the atmosphere and the depletion of the ozone layer; Melting of the ice caps leading to raised sea levels; Sea becoming warmer and more acidic; More extreme weather events: floods, tornados, hurricanes etc.
3. Loss (or extinction of) many species	3. Human activity causing loss of habitat or food source for many species.

CHAPTER 2
The industrial revolution

The most significant development, which is being linked to the environmental changes we are seeing in our world is the industrial revolution, so I will deal with it in some depth from an historical point of view, so we can all see how it came about and how it is still continuing.

Britain is considered to have been the birthplace of the industrial revolution which, historians say, took place during the period of 1760 to 1840. Before this, societies were mainly rural and the daily existence of small communities revolved around farming. Life was difficult, with the majority of people on low incomes, so many were malnourished and diseases were rife. People produced most of their own food, clothing, furniture and tools, with manufacturing (cottage industries) being carried out in homes or in small, rural shops, using hand tools or simple machines. The industrial revolution was to completely turn this around, having an impact on every family in the land and on their way of life.

Several factors contributed to Britain's role as the birthplace of the Industrial Revolution. It had great deposits of coal and iron ore, which proved essential for industrialisation and it was a politically stable society. At the time, it was also the world's leading colonial power, which meant that its colonies could serve as sources of raw materials, as well as a marketplace for manufactured goods. As demand for British goods grew, merchants needed better methods of production, which led to the rise of mechanization and the whole factory system.

One of the first inventions to spark the industrial revolution was in the textile industry: by the spinning "jenny", invented by an Englishman James Hargreaves in 1764. It was later improved on by others, and led to the power loom, which mechanised the process of weaving cloth, leading to the production of textiles on a wide-scale. Industrialisation of the textiles industry meant that some craftspeople were replaced by machines. This led to the Luddite Rebellion in 1811-1813, in which textile workers protested against the newly developed labour-economizing technologies which replaced them with less-skilled, low-waged labourers, leaving the craftsmen without work.

Industrialisation of the textile industry was followed soon after by the development of the iron industry. Englishman Abraham Darby discovered a cheaper, easier method to produce cast iron, using a coke-fuelled furnace and then, in the 1850s, British engineer Henry Bessemer developed the first inexpensive process for mass-producing steel. Both iron and steel became essential materials, used to make everything from appliances, tools and machines, to ships, buildings and infrastructure.

The steam engine was integral to the industrialisation process. In 1712, Englishman Thomas Newcomen had developed the first practical steam engine (which was used primarily to pump water out of mines) but, by the 1770s, Scottish inventor, James Watt, had improved on this and the steam engine went on to power machinery, locomotives and ships in the years that were to follow. Some say that the steam engine represented a second phase of the industrial revolution though many of these new technologies did overlap.

Fig.16: Image of the Industrial revolution in Britain

Coal mining became a major industry in the 19th century, as coal and/or coke was needed to power up the factories, as well as the engines running the railways and steamships.

The Industrial Revolution brought about a greater volume and variety of factory-produced goods and raised the standard of living for many people, particularly for the middle and upper classes. However, life for the poor and working classes continued to be difficult. Wages for factory workers were low and working conditions could be dangerous and monotonous. Unskilled workers had little job security and were easily replaceable. Children were part of the labour force, often working long hours and involved in hazardous tasks. In the early 1860s, one-fifth of the workers in Britain's textile industry were younger than 15.

Fig.17 Children working in a textiles factory (From: www.primaryhomeworkhelp.co.uk)

Fig.18: Young boys working as miners during the industrial revolution

From: http://historylearning.com/great-britain-1700-to-1900/indrevo/coal-mines-industrial-revolution/

Additionally, urban, industrialised areas were unable to keep pace with the flow of workers arriving from the countryside, resulting in inadequate, overcrowded housing and polluted, unsanitary living conditions in which disease was

44

rampant. However, conditions for Britain's working classes began to gradually improve by the later part of the 19th century, as the government instituted various labour reforms and workers gained the right to form trade unions.

The invention of the steam engine led to significant improvements in transport, from largely horse-drawn methods to the introduction of steam-powered engines for ships and railways. Steam powered cars first appeared in the late 19th century but these were to be replaced later by the, more popular, petrol driven engines.

The Motor Car

The history of the development of the motor car is well-known to us, as is the rise in the use of motor cars in the last 100 years, which has been phenomenal, with many households now being 2 or 3-car families, or even more. The thing that underlines this to me is the change in the road where I grew up. In the 1950's, there were no parked vehicles on this road and goods were often delivered with horse-drawn vehicles. When I last visited this street, in 2005, there were cars parked on both sides of the road, with room for only one vehicle to pass between them; woe betide if anything was coming the other way. Traffic jams are now a world-wide phenomenon, particularly in capital cities. My visits to Bangkok and Manila in 1994 were an eye-opener; in both of these cities, if you wanted to get anywhere by car and quickly, you had to leave home very early in the morning.

Fig.19 Takeover by the motor car

With petrol being a major culprit in contributing to carbon emissions, it would be expected that vehicles propelled by cleaner forms of energy would be starting to take over from petrol and diesel-driven vehicles but a chart published by Statista.com[17] shows that the total number of new vehicles registered in the UK has been on the increase and few of these use alternative fuels. The actual numbers for the UK in 2014 were:

		%
Petrol fuelled vehicles new registrations	1,184,409	47.8
Diesel fuelled vehicles new registrations	1,240,287	50.1
Alternative fuel vehicles new registrations	51,739	2.1

So, the small increase in the use of alternative fuel is minimised by the continuing dominance of petrol and diesel fuelled vehicles and the ever-increasing overall numbers of cars being used on the roads.

From another source (Society of Motor Manufacturers and Traders - SMMT)[18], I have obtained further information about the preferred type of alternatively-fuelled cars during 2014, compared with 2008 and 2011. The figure below shows that there is an increase in purchase of these vehicles over the six-year period, with a 25% increase between 2013 and 2014.

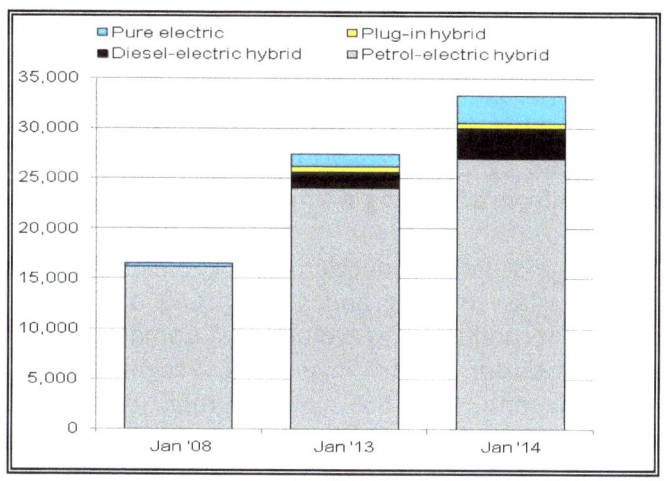

Fig.20: SMMT New car registrations for 12-month periods 2008, 2011 and 2014
Source: AIS 0207 235 7000

The continuing increase in vehicles of all kinds on the roads may be as a result of increases in the human population, or in an increased interest in driving by the developing countries of the world, as they try to catch up with the lifestyles of the developed countries. However, a recent initiative by Mexico City, to reduce the amount of pollution and smog in their capital city, has been to ban all vehicles from their roads for one day per week.

In the Netherlands in 2013, 1.4% of car sales were fully-electric vehicles and the Netherlands are currently second in the world (behind Norway) in adopting the highest number of fully electric plug-in vehicles[19]. Owners of these vehicles are already eligible for tax breaks and parking spots – a not surprising development in view of the vulnerability of this country to flooding as sea levels rise. A number of Dutch politicians are proposing the banning of gas and diesel-powered vehicles from 2025[20]. And Norway will completely ban all petrol-driven cars by 2025. Fig.20 shows that the UK is a long way from achieving a similar target.

The large-scale production of chemicals, then cement, glass making and gas lighting also began during the industrial revolution. Communication became easier with inventions, such as the telegraph and, in 1866, a telegraph cable was successfully laid across the Atlantic.

Thus Britain was the seat of these massive changes in industrialisation but it was not long before it spread, first to countries in Western Europe and then to America, being well established in these countries by the mid-19th century. By the early 20th century, America had become the world's leading industrial nation and remains so. Japan's industrial revolution began in about 1870 but other large eastern nations followed much later. China's was not until 1979 to 2000 and it still continues to escalate. India came under the East India Company at the time of the start of the Industrial Revolution in Britain, so there was some technological progress (such as the introduction of railways, canals, modern banks and postal system) but no significant advances during the 19th and 20th centuries, due to problems caused by some major famines and factional rivalries and wars, though India has been a major supplier of raw materials to Britain.

However, there are some people who believe that Britain's colonisation of India and the sequestration of its resources, set that country back years, leading to the destruction of many forests, loss of land rights and the subversion of its education and cultural traditions, especially associated with arts and science (W. Pereira and J. Seabrook, 1996[4,21].

'Follow Green Living'[22] talks about the Uttarakhand (flooding) disaster, which was caused by deforestation. The World Wildlife Fund has stated that every minute, forest area equivalent to 36 football fields is lost, along with 137 species of plants, animals and insects, which totals 50,000 species a year.

Fig 21: Indian weavers at the 1886 Indian and colonial exhibition in South Kensington © Victoria and Albert Museum, London

Fig.22: Deforestation in India

Some countries have still not become industrialised and continue to be mainly agrarian, rural or nomadic communities.

Other changes associated with the industrial revolution

During the industrial revolution there were changes in the economy, society and culture, perhaps some of the most significant changes in human history. It was much more than just a mechanization drive. It was also an epoch in European social history that characterized the transition from feudalism to capitalism and the development of the latter. So, there was a change from family-based economies, organised around and within agrarian communities, to an economy organised around a factory system, dependent on owners and managers, and on businesses and their productivity. The factory replaced the home as the centre of production. The industrialists running factories pressured governments to spend money on infrastructure (railways, roads, shipping etc.), to foster free trade between nations, and not to interfere with businesses and the way factories were run. This change in the focus of the economy will be discussed in more detail in chapter 7.

The industrial revolution also saw the rise of banks and industrial financiers. A stock exchange was established in London in the 1770s; the New York Stock Exchange was founded in the early 1790s. In 1776, Scottish social philosopher Adam Smith, who is regarded as the founder of modern economics, published "The Wealth of Nations."[23] In it, Smith promoted an economic system based on free enterprise, the private ownership of means of production, and lack of government interference. In the 21st century we have seen how the increasing power of banks has upset the balance of the economy, leading to vast profits for bankers at the expense of the average person.

So there have been many downsides to the industrial revolution, not the least of which has been the concomitant changes that have occurred to the earth's ecosystems, its biosphere, to global temperature and to the earth's climate.

The Industrial Revolution (IR) Continuum

Historians now say that the industrial revolution was followed by a second one, which continued from 1870 to 1914, with advances in technology, and a 3rd one later which included the digitisation of manufacturing and the internet and others are now saying that we are entering a 4th industrial revolution, marked by further advances in technology, which will fundamentally alter the way we live, work and relate to one another – included in this revolution will be advances in green technology. I personally don't think it is helpful to divide the industrial revolution into historical eras. This is because I believe that the chain of events the first industrial revolution initiated have continued to the present day. I call this process, which is still ongoing, the IR Continuum (i.e. the continuation of the first industrial revolution) and will use this name throughout the rest of this book.

Let's look at a few of the things that have been invented since the late nineteenth century which, along with the industrial revolution, have changed the face of this planet and had a large impact on our experience as human beings living here.

1. Electricity

The invention of electricity and the introduction of light bulbs by Edison in 1879 made a huge impact of the human way of life, as it extended the length of the day in which we could be active, from early morning until well into the evenings, as well as enabling people to work night shifts.

Fig 23: The development of the light bulb had a huge impact on society

The light bulb was followed by labour-saving devices, all powered by electricity; things for the home, such as washing machines and later dish-washers and the development of radio and television, as well as the motor car and other inventions described later in this chapter. Whilst the motor car is powered by a petrol or diesel engine, electricity is needed to maintain and circulate that power.

The problem with electricity of course is that, to generate it, we have been burning fossil fuels. And nuclear energy, now often promoted as a clean source of energy, is not the answer either as it has its own dangers from accidents (as in Chernobyl and Fukushima) as well as problems and dangers associated with disposal of nuclear waste. The present preferred means of generating energy are solar power, wind or water power but, as yet, they contribute to only a small proportion of electricity generation (see figure below, where renewable energy is marked as RE) and globally the proportion is even lower than that of the UK (see Fig. 25)[24].

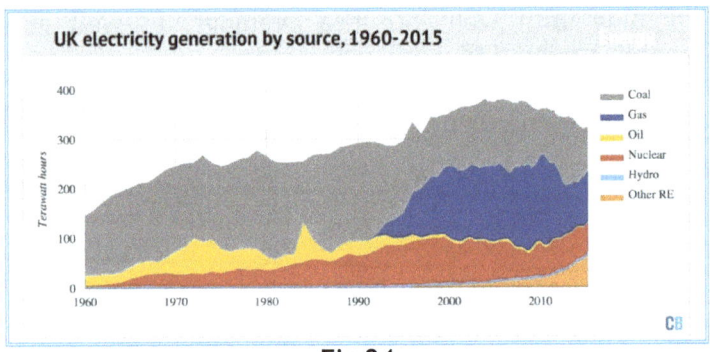

Fig.24

Figure 24 shows UK electricity generation by fuel type since 1960. In 2014, the electricity mix was 31% coal, 31% gas, 19% renewable and 18% nuclear. Chart by Carbon Brief using DECC data. From
www.carbonbrief.org and
http://www.carbonbrief.org/data-dashboard-energy-archive

And an interesting development in 2016 (also reported by Carbon Brief) is a 22% decline in coal use in the UK since 2014, coal now being replaced by renewables and nuclear power in the generation of electricity, as reported in The Guardian, 24th September 2015[25].

Changing to renewables is not a world-wide phenomenon, though. The biggest problem is that the global use of energy continues to increase; it has tripled since 1965, as shown in figure 25, with coal, gas and oil being the major energy sources.

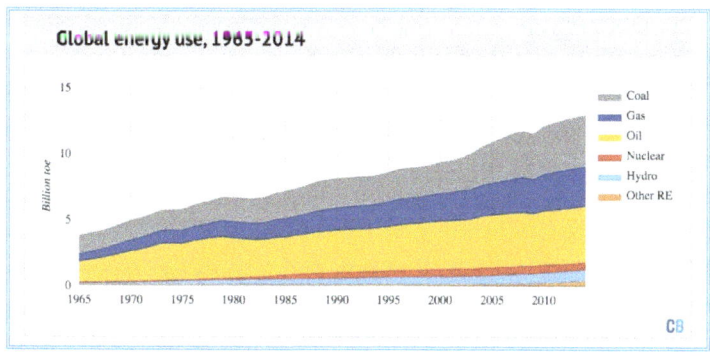

Fig.25 Global energy use by source, 1965-2014.
Source: BP Statistical review of World Energy 2015. Chart
by Carbon Brief: www.carbonbrief.org and
http://www.carbonbrief.org/data-dashboard-energy-
archive

Despite the small proportion of renewable energy
shown in the global graph, there is hope, as The
Guardian recently reported[26] that at least five
countries have shown initiatives to reduce reliance
on fossil fuels. For example: Uruguay gets 94.5% of
its electricity from renewables, due to a hefty
investment in wind, biomass and solar in recent
years. Costa Rica went for 94 consecutive days
earlier this year without using fossil fuel for energy
at all, thanks to a mix of 78% hydropower, 12%
geothermal and 10% wind. Iceland is able to tap
geothermal sources for 85% of its heating, which
with hydropower, enables this country to have 100%
of its electricity from renewables. Paraguay has one
huge hydropower dam at Itaipu, which supplies 90%
of its energy and Lesotho gets 100% of its electricity
from a cascade of dams that have enough spare
capacity to export power to South Africa.

2. Plastics

The first substance that could be described as plastic was Parkesine, produced by Parkes in 1862. It was highly flammable, so later versions followed, such as celluloid, bakelite, artificial silk, cellophane, polythene etc. The great advantage of plastics is that they can be moulded into any shape that is required and much of our life activities today are surrounded and influenced by plastics in one shape or another. The downside of plastics is that most are not biodegradable. So, the world now has many rubbish dumps, landfill sites and tips, all containing plastics, as well as other kinds of rubbish, the best example being that of Smoky Mountain near Manila. In Britain, we are rapidly running out of landfill sites in which to dispose of our rubbish. And we are told that our oceans are full of plastics, which damage marine life and wash up on beaches across the world.

Fig.26: An Indian boy walks by the Arabian Sea near Mumbai, piled with mainly plastic rubbish

A young Dutch student may have come up with a cheap solution with which to clear the oceans of plastic rubbish, using the pre-existing ocean currents[27] but this is not yet tried on a global scale and there would still be a problem of disposal, once the plastic is collected.

3. Weaponry

During the industrial revolution, there was a big development of new forms of weaponry, with hand-held weapons becoming ever more sophisticated and playing a major part in the First World War (1914-18). By the time of the Second World War (1939-45), there had been a development of bombs, as well as the aircraft to drop them on enemy targets. This culminated in 1945 with the dropping of the atomic bomb on the Japanese cities of Hiroshima and Nagasaki, killing 129,000 people immediately with many thousand others dying later as a result of nuclear fallout. Nuclear weapons have not been used in war since but are still owned by a number of countries, being used it is thought, as a deterrent to war.

Using weapons and bombs in war also has the effect of increasing carbon emissions. For example, it has been calculated that, during the Iraq war, the total carbon emissions per year, as a result of the war, were higher than the emissions of 139 countries put together. This issue is of such importance that I have devoted a whole chapter to it (Chapter 6).

4. Aviation

During the first half of the 20th century, there were huge advances in the manufacture of flying machines, first of all for early pioneers to fly over the great oceans but, later, to the development of passenger airlines, with some of the larger planes, such as the Airbus, now taking well over 500 passengers at a time.

Fig.27: The Airbus A380

In 1952, the first commercial jet flight took place and 24 years' later, Concorde began its fascinating history. Air travel has become so commonplace that it is now nothing special to fly to the other side of the world and back in a short space of time and to do it several times a year.

In her piece entitled "Counting the Cost", written for the New Era Network in 2005[28] (and downloadable from their website, Molly Scott Cato MEP, a green economist, gave some compelling statistics about carbon dioxide emissions related to the aviation industry, the expansion of which has been completely unregulated. Much of this increased usage of passenger airlines has been as a result of the expansion of tourism. In 1990, CO_2 emissions from aircraft accounted for about 2.4% of total emissions – they are projected to grow by another 3-7% by 2050 to approximately 10% of all emissions. The entire transportation system accounts for about 25% of emissions. Global tourism increased from 8.5 million people in 1970 to 56.8 million in 2000. So, the current obsession with taking regular holidays (or mini-breaks) and flying around the world to some remote destination is a major contribution to the problems of global warming and climate change. Yet, when people come into an unexpected financial windfall, the majority of them plan to use it first to take an overseas holiday. The effects of this on global warming are rarely thought of. The airline industry is certainly not going to advertise their impact on climate change, for it might mean the loss of their business.

5. Electronics, Space and the digital revolution
After the Second World War, we entered into an electronic era, the Space Age, satellite technology and the digital revolution, which began with the invention of the transistor in 1947, followed by computers, hard disks, microchips and microprocessors, recording devices for music and video/film, CDs, DVDs, iPODs, hand-held devices (calculators, 1972; mobile phones, 1983), computer games, smaller and smaller computers, the internet, computer software, Facebook, YouTube, Twitter, smart phones, robots etc.

The development of space travel will be discussed further in the next chapter.

The Consumer Culture

The digital age has revolutionised the human way of life on a global scale, with vast advances in communication, which could never have been anticipated when the telegraph was first invented in the mid-1800s. The downside of it is that some of the devices invented are superseded very rapidly by improved versions, leading to a throw-away culture, as people try to obtain the latest version of the devices they treasure. All of this, of course, feeds into the escalation of the IR Continuum.

Whilst a significant number of people adhere to the consumer culture, wishing to have the latest invention in line with their friends and colleagues, there are those who are deeply concerned about it. In fact, when I first talked about writing this book to some of my friends, the throw-away culture that we live in was the first thing that sprang to their minds. Not only is it damaging to the planet, feeding into the IR Continuum and the accumulation of discarded items, it is also bound up with trading patterns and an obsession with economic growth, as discussed in chapters 4 and 7.

Who are the worst polluters?

It is the industrialised countries that have contributed most to carbon emissions, though the whole world feels the effect of this. And amongst the industrialised nations, some emit more per head than others. Damon Matthews from Montreal in Canada has calculated climate debts for each country related to their population size. He sees those who pollute more than their fair share (i.e. above the global average), as being in climate debt. From these calculations, the US leads the list by a long way, with the greatest climate debt, Russia is second and Japan third; the UK is the 6th worst polluter in the world[29]. Other ways of presenting the data show the UK in first place (because we have been industrialised for longer).

However, in terms of individuals, the richest people in the world contribute to 85% of carbon emissions (see also in Chapter 5).

Fig.28: The different ages of man up to the present computer generation

INTERCONNECTIONS WITH THE INDUSTRIAL REVOLUTION DISCUSSED IN THIS CHAPTER

CONNECTION	AS SHOWN BY
1. Industrial Revolution	1. Climate change and global warming due to increased emissions of carbon dioxide and other pollutants.
2. Industrial Revolution	2. Economic changes as agrarian economies move to business economies.
3. Industrial Revolution	3. Social changes, based on feudal systems, based on land owners and vassals (who provide services for tenure), to business-based (factory) employers with employees.

4. Industrial Revolution and its continuum (IR Continuum)	4. Increase in standard of living and lifestyles, leading to the ability to purchase more products from industry and a throw-away culture.
5. Industrial Revolution and its continuum	5. Increase in the human population of the world.
6. Industrial Revolution and its continuum	6. A movement of people from rural areas to cities in search of work.
7. IR Continuum	7. Technological changes, leading to better communication, the conquest of space, satellite technology and the digital age.
8. Industrial Revolution	8. Changes to trading systems from local exchange and barter to global trading systems and the power of embargos to trade.
9. Industrial Revolution	9. The rise of banks and stock exchanges.

CHAPTER 3
Human inventiveness and the concepts of progress and freedom

Human inventiveness

The "IR" continuum of development, which has continued unrelentingly since the industrial revolution began, demonstrates the huge human potential to invent, explore, manufacture, develop, innovate and improve - something that we can be proud of as the species which has come to dominate this planet. Human achievements are extraordinary and the potential to continue in a similar vein is tempting and attractive, each generation wanting to better the one before and each nation wanting to do better than its rivals. But somehow, we need to differentiate between natural pride in our achievements, curiosity, the desire to invent more and more complex gadgets and **the need for responsibility**. Otherwise human inventiveness becomes a continuum of its own and a weapon of our own destruction.

One of the best examples of this is, perhaps, the situation we find ourselves in, as a result of space travel and satellite technology. The picture below, recently featured in a BBC Horizon programme, in which we were informed that there are 22,000 large objects circulating the planet, each travelling (hurtling) at 17,000mph. The result of collisions could be catastrophic, yet there is currently no law governing space operations. At the moment about 120 new satellites are launched every year and are used for all kinds of purposes, from GPS systems to TV programmes linking events happening across the world and drone warfare. Some of them are just debris remaining from previous space missions. Only recently part of an American space rocket washed up in the Isles of Scilly.

Fig. 29: Picture showing Space junk circling the earth
(European Space Agency) See also the animated video at
http://www.rigb.org/docs/debris/ from the Royal
Institution

The idea of launching rockets, space stations and satellites into space is exciting for us as a species but, over the last 40-50 years, it has been carried out with no forethought, no cohesion, inadequate co-operation between nations and with no real plan to mop up the debris it has created. In fact, in the early days, it was carried out in competition between nations, America and the Soviet Union each wanting to be the first to put a man into space or on the moon. This is only one issue where there is a need for more, and better, co-operation between nations. In a way, the image is a dramatic reminder of how man, left to his own devices, has messed with the planet almost irrevocably.

Fig 30: The launch of a space rocket

The Concept of Progress

Everything that has happened since the beginning of the industrial revolution has been seen as progress. But is this so-called progress really progress at all or is it a hedonistic route leading to the destruction of the planet? This is assuming that we do not destroy each other in a fight for supremacy in the process. One of the voices raised against those who warn that we need to live more simply, is that we cannot ignore progress and return to the times before the industrial revolution, which were not as comfortable as the times we live in today. It is seen as a backward step. And I am definitely not advocating that we return to the way things were before the industrial revolution.

So, let's have a look at this idea of progress and what it really means. Perhaps it means different things to different people. Perhaps it is being used to justify practices that are dangerous to the planet and to our future existence, for many have benefited financially from the relentless IR Continuum.

The word "progress" itself gives a feeling of dynamism and of continuous advances and improvements. But, if these advances are not beneficial to the global population as a whole, and its many species, then they cannot be described as progress at all but are more accurately described as "destructiveness" or "dissolution". I would therefore like to re-define the word "progress" to: *making advances that benefit society and the global population"*. With this new definition, we could still include advances in technology, such as the development of forms of renewable energy and some medical advances, as progress. But everything else should not be deemed as progress at all.

Freedom of Choice

Along with "progress" another word that is attractive to us as a species is the concept of "freedom" or "free-will". Individuals like to be able to have a choice about what they do and to have the freedom to engage in whatever it is that fascinates them, whether it be space technology, weaponry, driving fast cars, inventing new machines or just taking a holiday abroad. But freedom has to be accompanied by **responsibility**, or it just becomes selfishness. Any suggestion that we should curb our lifestyles is opposed rigorously. Yet unpopular legislation that has happened in the past (like seat-belt legislation, MOT testing of vehicles and the rationing of food after the Second World War) has eventually led to greater safety and more responsible behaviour. In an effort to reduce the number of plastic carrier bags littering the world, some supermarkets have recently started to charge for these. A good idea but I have been gobsmacked at the negative response to this in some quarters.

Two Opposing Dynamics

I believe that there are two dynamics in operation here. On the one hand there is the wondrous beauty of the world that we live in, a world where everything fits harmoniously together in the cycles, webs and relationships described in chapter 1. And, on the other hand, there are the mind-boggling achievements of the human race, developed through our special intelligence, which sets us apart from other species. The two dynamics are juxtaposed but, as yet, not in harmony. Can we, as intelligent human beings, find a way forward in which our inventiveness and inquisitiveness and (yes) greed, does not destroy the beautiful world in which we live, before we end up destroying ourselves and the habitats we, and the creatures we share this planet with, live in?

The two opposing dynamics in juxtaposition

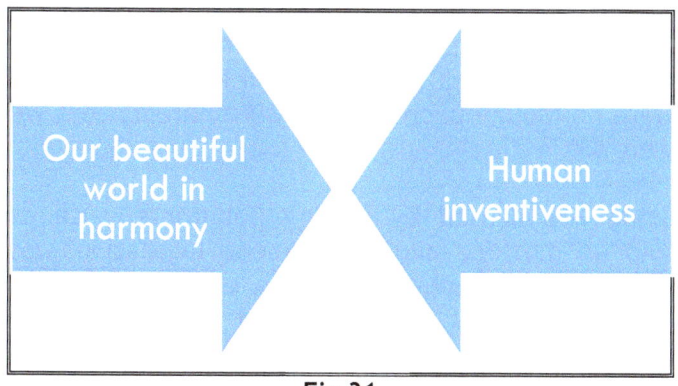

Fig.31

Britain's role in saving the Planet

Furthermore, since Britain was the country to introduce industrialisation to the world, I believe it should also play a major leadership role in finding ways to reduce the damaging effects of carbon emissions and other pollutants (methane, chlorofluorocarbons, nitrous oxide, black carbon particulates, nuclear materials) produced by industrial economies. And, in a similar vein, perhaps the responsibility of clearing up space junk should rest with the space-pioneering countries, America and Russia and, to a lesser extent, Europe.

During the 1940s-50s, one of the problems faced by Britain was *smog* – a thick, dense fog that descended on our cities, making breathing difficult and reducing the ability to see where you were going. A powerful memory from my childhood was a time when smog had descended so thickly that we could not find our way home; my sister and I therefore decided (perhaps unwisely) to take a bus; the bus did eventually arrive but the smog was so thick that the driver could not see the kerb and the bus conductor had to walk alongside the bus with a torch, to help the driver to see the way ahead. It took us hours to get home and, in those days, there were no mobile phones to reassure parents that we were OK.

Smog was caused by factory pollutants, such as sulphur dioxide, and government legislation was brought in to limit the emission of such chemicals and, as a result, smog has not been the same problem that it was during my childhood. That particular era demonstrates that it is possible to reduce the harmful effects of industrialisation. However, we still see a haze of pollutants in the skies over many of the major cities of the world. It was observing the cloud of haze over many of the cities of the world during my 1994 world trip that made me realise that one day I would have to raise awareness about it. It has taken me 22 years to take such action but I do hope that the action required to save the planet does not take another 22 years.

Fig.32: People in China wearing masks due to the urban pollution there

Interestingly, I am not the only person to use the smog era as an example of what can be done if the motivation is there, through the British legislation at that time to prevent further smog. Prof. Paul Rogers, in his ORG Special briefing to the Oxford Research Group[30] and his lecture at the Imperial War Museum (2012) has also cited it, together with three other examples of actions that can be introduced to alleviate the worst effects of human activity. The following are four of his examples:

- Municipal engineers like Sir Joseph Bazelgette were already working on plans for proper sewage disposal in the squalid and cholera-ridden London of the 1850s, but it took the "Great Stink" of the 1858 summer to prompt sustained and effective action, with London leading the way for many other cities and resulting in sustained improvements in health.

- A century later, 4,000 people died in 1952 in the four-day "Great Smog of London" but this prompted radical improvements in air pollution control across Britain that were already being called for.

- Atmospheric chemists and the UN Environment Programme were already pointing to the dangerous effect of CFC pollutants on the world's ozone layer in the late 1970s, and, partly because of this, the discovery of the extent of the Antarctic "ozone hole" in 1983 prompted a rapid international response, with the global Montreal Convention signed just four years later.

- Even at the height of the Cold War, the shock of the Cuban Missile Crisis of 1962 was a major factor in leading the US and the Soviet Union to a welcome process of trust building and some key agreements, including the Partial Test Ban Treaty, as well as leading to a political climate that helped bring about the Non-Proliferation Treaty

The move of populations throughout the world to cities has also led to urban sprawl and cities becoming more and more industrialised. Pollution of the atmosphere around cities is very evident. A recent article in the Financial Times[31], reporting on studies at Kings College London, claimed that up to 9,400 Londoners per year die prematurely because of breathing in pollutants commonly found in fumes from diesel trucks, buses and cars. Thus Londoners are more likely to be killed by the air they breathe than in a car accident. During 2010, more than 3,000 people were admitted to hospital with breathing and heart problems related to air pollution.

So, in summary then, human inventiveness, whilst being a remarkable feature of the human species, can also have its downside. Not all inventiveness should be seen as progress. We need to develop as a species a new quality – that of being able to assess what destruction unregulated-inventiveness might bring and acting responsibly as a result. We do still have a beautiful world, even though it is in danger. Can I add another dynamic to the picture? Is this achievable?

Fig.33: The three dynamics in balance

CHAPTER 1
Trading Systems, Deficits and the Concept of Growth

International trade has become so much a part of our lives that there is a tendency to take it for granted, as a normal and essential part of modern society and that of the countries of the world with which we trade. Politicians particularly focus on it, as it is seen as a means of balancing the economy; they particularly encourage the export of British goods and turn a blind eye to all the stuff that we import.

The industrial revolution and its continuum and the development of trading systems

Historically though, trading systems as we know them today were first developed alongside the Industrial Revolution. And again, the UK was a forerunner in developing these new trading systems, as they sold the goods produced in their factories to other countries across the world, particularly to members of the British Empire, such as through the East India Company in India. This change from the local exchange of goods to the export of goods across continents and the world has had such a great impact that its influence now affects, and influences, the whole world's economy. The nations of the world have become so inter-connected through trade that, if one country goes through economic difficulties, then all the others are affected by it too. Because of the strong link between trading and the industrial revolution and its continuum, I have to consider it, and its effects, as one of the major interconnections that has led us globally to the situation in which the future of our planet is at risk. **Indeed, I believe that free trade is at the centre of it all**.

The Industrial Revolution ended more than a century ago but the effects of it, the trading systems that were developed alongside it and the IR Continuum, still have a growing global impact.

The effect of the IR Continuum on global trading systems has seen the rise of multi-national companies (mostly of American origin), not only trading with other countries but also setting up business abroad, in order to cut costs, employ cheaper labour and to avoid national tax tariffs. It is not unusual now to see MacDonalds, Kentucky Fried Chicken, Monsanto and other multi-national outlets in most capitals of the world. This is sad because the setting up of food and clothing outlets selling goods that promote the American way of life has the effect of damaging indigenous cultures and their traditions.

We also see locally produced goods transported across oceans and continents in order to trade with partner countries many thousands of miles away. In the UK, for example, we import apples from New Zealand and Chile, fruit from South Africa, fish from Japan and Argentina, clothing and digital goods from the Far East, vehicles from Europe and so on. The invention of the refrigerator has played its part in preventing perishable goods from decomposing whilst in transit.

Fig. 34 A multi-national outlet for the USA in Japan

Changes in trading patterns across the world since the industrial revolution can also be contentious. For example, when I lived in Australia during the early 60s, the UK was considering whether it would join the European Common Market (now the EU). This was very unpopular with Australians, as they had a special trading relationship with the UK, as part of the British Commonwealth. However, Britain **did** join the EU and so Australia had to develop other markets, closer to home, and were able to survive this change. But the resentment it caused in some Australians towards the EU, and the British, is still present today, as seen by the anti-EU stories constantly being peddled to the UK population, through the Australian-owned media magnates.

There has been a big change in Britain's trading patterns as, during the 1940s-50s, about 40% of our trade was with Commonwealth countries but this is now down to 10%, as the EU has become our major market.

Large Companies and Climate Change Denial

The largest company in the world, ExxonMobil, produces oil and gas and a recent article by Shannon Hall, in Scientific American[32] reports that this company was aware of climate change as early as 1977, before it became a public issue. The company then spent decades refusing to publicly acknowledge climate change and even promoted climate misinformation. Hall likens this approach to the lies spread by the tobacco industry regarding the health risks of smoking. Exxon became a leader in campaigns of confusion and helped create a *Global Climate Coalition* to question the scientific basis for concern about climate change. It also lobbied to prevent the USA from signing the Kyoto Protocol in 1998 (to control greenhouse gases), also influencing other countries, such as China and India, not to sign as well. It has spent $30 million on think tanks that promote climate denial, according to Greenpeace. Hall's article provides data that suggests that half of the greenhouse gases in our atmosphere have been released since 1988. If ExxonMobil had been upfront about the issue in those early years, there could have been so much more progress on climate change than there has been. The company obviously had vested interests in opposing the scientific evidence but they now have a lot to answer for. And there are now rumours that Shell is under investigation for doing a similar thing.

It has recently been reported that one of the major American charitable foundations (Rockefeller Family Fund) has announced that it will cease to invest its funds in fossil fuels and, in doing so, made the following statement: *"We would be remiss if we failed to focus on what we believe to be the morally reprehensible conduct on the part of ExxonMobil"*.[33]

ExxonMobil would have better spent their $30 million researching into new forms of renewable energy.

Table 3 shows that, in 2009, there were three energy companies amongst the ten largest companies in the world, but by 2015, this had risen to be five in the top six, all of them in the petrol refining business. However, despite their dominance in world markets, energy companies have much to lose, once the issue of carbon emissions is properly dealt with by global agreements to reduce them.

TABLE 3

THE 25 LARGEST COMPANIES IN THE WORLD in 2009
From:
http://bespokeinvest.typepad.com/bespoke/2009/04/larges
t-companies-in-the-world.html

Ticker	Country	Company	Sector	Price
XOM	United States	Exxon Mobil Corp	Energy	65.95
601857	China	PetroChina Co Ltd	Energy	11.72
601398	China	Indu & Comm Bank of China	Financials	3.99
WMT	United States	Wal-Mart Stores Inc	Cons Stap	48.42
MSFT	United States	Microsoft Corp	Technology	20.53
941	Hong Kong	China Mobile Ltd	Telecom	67.25
T	United States	AT&T Inc	Telecom	25.15
PG	United States	Proctor & Gamble Co	Cons Stap	49.79
RDSA	Netherlands	Royal Dutch Shell PLC	Energy	1557.00
JNJ	United States	Johnson & Johnson	Health Care	50.96
BRK/A	United States	Berkshire Hathaway Inc	Financials	90240.00
7203	Japan	Toyota Motor Corp	Cons Disc	3790.00
IBM	United States	IBM	Technology	100.24
PETR3	Brazil	Petroleo Brasileiro SA	Energy	35.59

601939	China	China Construction Bank	Financials	4.35
CVX	United States	Chevron Group	Energy	65.46
BP	Britain	BP PLC	Energy	483.25
GE	United States	General Electric Co	Industrials	12.11
NESN	Switzerland	Nestle SA	Cons Stap	38.16
JPM	United States	JPMorgan Chase & Co	Financials	33.58
BHP	Australia	BHP Billiton Ltd	Materials	32.41
GOOG	United States	Google Inc	Technology	385.41
FP	France	Total SA	Energy	38.25
HSBA	Britain	HSBC Holdings PLC	Financials	466.50
601988	China	Bank of China Ltd	Financials	3.46

Footnote: By 2015, the top ten companies listed by Fortune (http://fortune.com/global500/) had changed to:
1. Walmart; 2. Sinopec Group, China; 3. Royal Dutch Shell; 4. China National Petroleum; 5. Exxon Mobil Corp; 6. BP; 7. State Grid, China; 8. Volkswagen; 9. Toyota, Germany;
10. Glencore, Switzerland (mining, crude oil production). Thus, in the six years from 2009 to 2015, there had been a change to increasing numbers of large companies in the petroleum refining business to five of the top six companies (or 11 of the top 25) in 2015.

Carbon Majors – the companies who emit the most greenhouse gases

90 carbon majors have been identified as being the major emitters of the greenhouse gases that are primary drivers of climate change. Since 1751, they have produced 65% of the world's total industrial carbon dioxide emissions according to a study by Richard Heede of the Climate Accountability Institute[34]. The 90 majors include 50 private companies, 31 state-owned companies and 9 nations. Twenty-one are based in the US, 17 in Europe (five in the UK), six in Canada, two in Russia and one each in Australia, Japan, Mexico and South Africa. Of the state-owned companies, Saudi Aramco has the highest emissions, followed by Gazprom (Russia), National Iranian Oil Company, Pemex (Mexico) and British Coal. The top 10 carbon majors are:

Chevron USA, ExxonMobil USA, Saudi Aramco Saudi Arabia, British Petroleum (BP) UK, Gazprom Russian Federation, Royal Dutch Shell, National Iranian Oil Company Iran, Pemex Mexico, British Coal Corporation UK and ConocoPhilips USA. For full details of these companies, and where they rank, are given by Greenpeace[35].

Last September Greenpeace Philippines were so concerned about the devastation caused in their country by a major typhoon, that they filed a human rights complaint to the Commission of Human Rights, against the 50 largest multi-national private companies[36].

The Volkswagen deception

ExxonMobil has not been the only large corporation to deceive the public on the issue of carbon emissions. Just recently, it has come to light that the large German car-manufacturing company, Volkswagen, has tried to avoid green regulations and tests by fitting its cars with devices to cheat the emissions tests carried out on vehicles. The scandal has resulted in Volkswagen shares falling by 40%. This deception is akin to the deception propagated by ExxonMobil, described earlier, where large and successful companies have used their trading links to make money for themselves at the expense of the health of the planet. One wonders how many more companies will come to light which are carrying out similar deceptions for selfish reasons.

Earlier this year, a new independent organisation was set up in London (InfluenceMap.org), to map, analyse and score the extent to which corporations are influencing climate change policy. An article in ExaroNews[37] published in 2015, reported that research from InfluenceMap has demonstrated that car manufacturers (especially those in Germany) have been lobbying strongly against climate-change policy, especially those who have made little progress in complying with future standards for emissions of CO_2 in the EU and US. The InfluenceMap article ranks car makers according to their compliance with the 2020 standard on emissions, with Nissan coming top, followed by Honda, Renault and Peugeot. According to the report, the world's 12 biggest car manufacturers would be facing fines of $35.7 billion if the 2020 rules on emissions were to be applied now, with Volkswagen paying more than any of them, at $9.5 billion. Car manufacturer Mercedes-Benz has admitted that meeting the 2020 emission standards poses a technological strain (also reported in ExaroNews). One wonders why none of them have acted sooner to develop greener cars, as some of the Japanese manufacturers have done.

Trade and Competition 1

The problem is that trading evokes a competitive spirit, even in the largest and most affluent companies, and the temptation to cheat can be persuasive. As well as the deceptions already mentioned, there has been the development of parallel economies, in which companies try to evade taxes and tariffs by investing their profits in offshore accounts. There are many people throughout the world who try to avoid national taxes by setting up their own parallel economies. They contribute to an underground economy or "black market", which is a market consisting of all commerce on which applicable taxes and/or regulations of trade are being avoided. It includes many multi-national businesses, as well as those involved in the growing and selling of illegal drugs.

Because trading has become an endemic part of the global economy, embargos on goods are often used as powerful political weapons to bring other countries "into line". Examples of this are the embargos on South African goods during the apartheid era and that currently being imposed on Russia because of its occupation of the Crimean region of the Ukraine.

The competition for markets associated with trade has far-reaching effects across the globe. Politicians talk about it as being a vital part of the economy and in so doing, they encourage this competitive spirit. Its linkages into the economy and how trade-associated competition is making global warming and climate change worse, will be discussed later in this chapter and in chapter 7.

The whole trading scenario reaches into many aspects of life and plays just as important a role in the development of climate change, as the industrial revolution has done.

OIL

Oil has also come to dominate global trading systems, with prices being hiked by the oil-producing countries, with non-oil-producing countries being held to ransom. Most governments fear that having no access to oil will impair their ability to manufacture and to trade, and thus impact on their national economies. The fear of losing access to oil has had a huge impact on national decision-making and the willingness to go to war to wipe out regimes who have large oil resources and who are not friendly to the western world. All of these fears, and the actions associated with them, are futile really because, if we are to save the planet, we need to stop using oil and other fossil fuels, by leaving them in the ground, and to replace them with renewable forms of energy.

Perhaps ExxonMobil and BP and other oil producing companies still need to learn this.

Fig. 35 An oil well

Further details about the movement of oil around the world (in terms of imports and exports) are shown on the Carbon Brief website[38], which appears to show that exports of oil were still increasing in 2014, compared with 2004.

At present, oil-producing countries have the upper hand but I do not see this as lasting, as there is a move to using non-carbon-emitting forms of energy, such as solar panels and wind, tidal and water-based energy. This could completely change the whole dynamic of global trading. If they seize the opportunity, some African countries in Saharan and sub-Saharan regions, could move from being poverty-bound regions, to replacing the oil-producing countries in the pecking order, by becoming leaders in producing and supplying cleaner forms of energy, such as solar power. Chile has already made a start by building a "farm" of solar panels in a desert area; this already supplies enough energy for one of their largest cities.

Fig. 36 Solar power farm in Chile

The trend towards renewable forms of energy has put some of the multi-national energy companies into a panic, as they search frenetically for oil and/or gas in more and more remote places, such as the Arctic.

There is a saddening history of how oil has damaged the environment and some animal and bird species, through oil slicks and spillages, yet the competitive urge to find new places to drill for oil and other gases continues unabated. The following three photographs show some of the consequences of oil spillage.

Fig. 37

Fig. 38

Fig.39

Fracking

Another area of concern is the new practice of fracking where licences have already been obtained to carry out this practice, which releases natural gas from under the ground in areas very close people's homes. Further information and an interactive map of the areas of the UK and Ireland affected by this can be found at the website: http://frack-off.org.uk/extreme-energy-fullscreen/.

News stories from Canada and America suggest that fracking there is linked to significant earthquakes.

Market Economies

The major change in trading systems across the world, since before the industrial revolution, has impacted substantially on the way of life and the economies of most nations of the world, so that whole economies are now based on trading patterns, potential markets and import/export ratios. Indeed, the description of a *market economy* is considered by some to be a progressive form of government. It is based on the concept of demand and supply, where governments encourage those companies in their trade who are meeting an overseas demand for their goods. The income they receive from overseas is seen to help the balance of payments and to bring about economic growth.

What a market economy fails to do is to analyse, and meet the needs of, its own people, especially those who are in poverty, with no goods to sell. The excuse for failing to help those in most poverty is that there will be a *trickle-down* effect; in reality this rarely happens.

What does happen is that the rich get richer at the expense of the poor.

Market economies are based on the encouragement of *free trade*, which is thought by 93% of economists to be a good thing (Ian Fletcher (2010)[39] but, as argued by Fletcher, it has led to a situation where some developed nations have huge trade gaps, or deficits, Britain being one of them. This has occurred mainly because some of the developing nations pay much lower wages to their industrial workers and can therefore produce and sell their goods at more competitive prices than those of the developed nations. In 2014 the trade deficit of the U.S.A. was $508,324 billion. Fletcher makes a case for rethinking and reforming current trade policies, by debunking some of the cherished assumptions held by mainstream economists. In the UK, the trade deficit for manufactured goods is higher than that of most other European countries but, in the past, politicians have worked to reduce the deficit by implementing austerity measures, rather than by rethinking our trade policies altogether, introducing localisation policies and making the reduction of carbon emissions a priority.

The UK Office for National Statistics (ONS) provides data which shows that the balance of trade in goods in the UK has shown a deficit in all but six years since 1900. They recorded net surpluses in the years 1980 to 1982, largely as a result of growth in exports of North Sea oil. Since then, however, the trade in goods account has remained in deficit (see Figure 40).

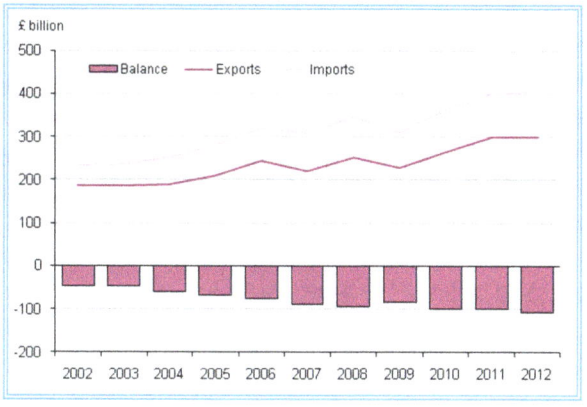

Fig.40
The trade deficit in the UK – from the Office of National Statistics

Figure 41 shows that Britain's trade in services is doing much better than its trade in goods.

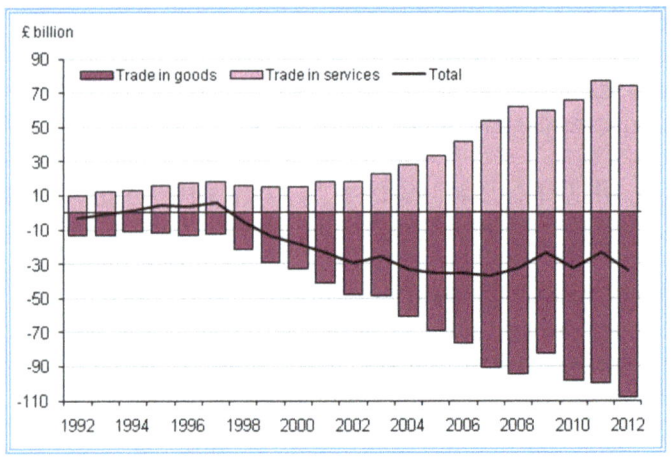

Fig.41
From the Office of National Statistics

The trade deficit also impacts on crops and foodstuffs produced by our farmers. In 2002, Dr Caroline Lucas, a Green MEP, wrote a report[40] entitled "Stopping the Great Food Swap: Relocalising Europe's Food Supply". It was based on background research and support provided by Andy Jones and Vicki Hird of Sustain and from Colin Hines, author of "Localisation: a Global Manifesto, published in 2000[41].

Lucas's report provides some astonishing data:

- The UK imports 61,400 tonnes of poultry meat <u>from</u> the Netherlands and, in the same year, exports 33,100 tonnes of poultry meat <u>to</u> the Netherlands;
- The UK imports 240,000 tonnes of pork and 125,000 tonnes of lamb while exporting 195,000 tonnes of pork and 102,000 tonnes of lamb;

- In the UK in 1997, 125 million litres of milk was imported and 270 million litres exported;
- In 1996, the UK imported 434,000 tonnes of apples, 202,000 tonnes of which came from outside the EU. Over 60% of UK apple orchards have been lost since 1970.

Thus, we are importing more agricultural goods than we actually export, and importing goods which we produce ourselves, yet our own farmers struggle to make an income. I have also come across figures which show that 46% of the food we eat is imported.

The report stated that trade-related transportation is one of the fastest-growing sources of greenhouse gas emissions and is therefore significant in terms of climate change.

In 2011, Rianne ten Veen, of GreenCreation, updated the Lucas report, providing more recent data, with three case studies on meat, milk and fruit, for the Counting the Costs series of reports[42].

The EU Common Agricultural Policy has been accused of creating a situation in which damage is caused to the environment and to rural livelihoods, by encouraging larger, more intensive farms at the expense of smaller, more sustainable ones and leading to the inhumane treatment of farm animals. There is evidence that the transport of livestock and meat across Europe has led to diseases such as Foot and Mouth Disease and BSE being passed from one country to another. The system has led to an absurd situation, which rewards a few, very wealthy farmers, the supermarkets and multinational food companies at the expense of small and medium-scale farmers. It makes no economic sense.

Further data is available in the report, which concludes that this destructive **globalisation** needs to be replaced with a **localisation** that protects and rebuilds local economies across the world.

The organisation, Local Futures, has recently released a 16-page action paper, entitled *Climate Change or System Change?*[43] which argues that globalisation (the deregulation of trade and finance through an ongoing series of "free trade" treaties) is the driving force behind climate change. The document makes the case for an international move towards *localisation* and provides a list of the pro's and con's for both systems, showing that the advantages of localisation far outweigh the advantages of globalisation. It provides evidence to demonstrate that globalisation:

- Promotes unnecessary transport;
- Promotes rampant consumerism;
- Is making the food system a major climate-changer;
- Replaces human labour with energy-intensive technologies;
- Promotes energy-intensive urbanisation.

A recent book by Colin Tudge[44] proposes a complete rethink of our approaches to farming, through "enlightened agriculture", without wrecking the rest of the world.

Economic Growth

Economic growth is defined as an increase in the capacity of an economy to produce goods and services, compared from one period of time to another. It is the long-term expansion of the productive potential of an economy. The problem with this is that this type of growth (as with so-called progress) is dependent upon relying on producing more and more manufactured goods and finding overseas markets to sell them. It all feeds into the IR Continuum, thus adding to further carbon emissions.

Growth is seen as a good thing by economists and politicians but, as with "progress", it can't be good if it is adding to carbon emissions and the destruction of the planet. At present, success in national economies is measured using an index called the GDP (gross domestic product). At the time of writing the growth in the GDP in the UK was 0.5% and, in the USA it was 1.5%.

In his book, "*The Growth Illusion: how economic growth has enriched the few, impoverished the many, and endangered the planet*" (1999), Richard Douthwaite[5,45] sets out how a capitalist system can be redirected to fulfil society's hopes by restructuring economies to be based on local rather than global imperatives. Some of his ideas will be looked at further in a later chapter.

Social Businesses

The Nobel laureate, Muhamad Yunus has promoted the concept of social businesses, which are businesses with social objectives (*Creating a world without poverty:* by Muhammad Yunus, 2007)[46]. He believes that we need to recognize the real human being and his or her multifaceted desires. In order to do that, we need a new type of business that pursues goals other than making personal profit - a business that is totally dedicated to solving social and environmental problems. He gives three examples of social businesses:

- One that manufactures and sells high-quality, nutritious food products at very low prices to a targeted market of poor and underfed children;
- A social business that develops renewable-energy systems and sells them at reasonable prices to rural communities that otherwise can't afford access to energy;
- A social business that recycles garbage, sewage, and other waste products that would otherwise generate pollution in poor or politically powerless neighborhoods.

It may be owned by one or more individuals, either as a sole proprietorship or a partnership, or by one or more investors, who pool their money to fund the social business and hire professional managers to run it.

A social business might be defined as a *non-loss, non-dividend business*. Rather than being passed on to investors, the surplus generated by the social business is reinvested in the business. Ultimately, it is passed on to the target group of beneficiaries in such forms as lower prices, better service, and greater accessibility. Not only does the investor get his money back, he still remains an owner of the company and decides its future course of action.

It is not known whether a social business feeds into the IR continuum as much as traditional businesses do but, because there are social and/or environmental objectives, one suspects that the carbon footprint will be much reduced because those who run the business are not there to make profit for themselves but to improve society. The Fair Trade movement also has social objectives.

The Organization for Economic Co-operation and Development

The OECD is a forum where the governments of 34 democracies with market economies work with each other, as well as with more than 70 non-member economies to promote economic growth, prosperity, and sustainable development.

In recent years there has been an OECD move to start measuring economies according to their **green growth**. In June 2009, ministers from these 34 countries with market economies signed a Green Growth Declaration[47], declaring that they will: **"Strengthen their efforts to pursue green growth strategies as part of their responses to the crisis and beyond, acknowledging that green and growth can go hand-in-hand."** They endorsed a mandate for the OECD to develop a Green Growth Strategy, bringing together economic, environmental, social, technological, and development aspects into a comprehensive framework. The Strategy was published in 2011 and formed part of the OECD contributions to the Rio+20 Conference in June 2012.

The strategy identified the following as being the most polluting industries with the greatest CO_2 emissions:

- Air transport;
- Water transport;
- Electricity, gas and water;
- Coke, refined petrol and nuclear fuel;
- Land transport;
- Basic metals;
- Non-metallic mineral products.

The document outlines ways to achieve international co-operation on the strategy and ways to monitor green progress. It is a significant document[47].

I would support the introduction of a new measure – a green GDP – which assesses only productivity associated with products which do not add to the total global emissions of CO_2 and other pollutants. Thus countries' outputs could be compared using both metrics:

- The normal GDP
- The green GDP

The OECD suggestion of monitoring the green GDP would give incentives to nations to lower their carbon emissions and to focus on developing products which run on clean energy or which can be manufactured with minimal emissions.

Another form of trading of the last few decades is in world currencies and commodities. National currencies vary from day-to-day, according to the world economic situation, and some people speculate in buying and selling currencies, like a kind of international casino. It is a form of risk that titillates the human need for excitement and intellectual entertainment, as does speculation on stock markets and commodities. But it can also help an individual to make money at the expense of some countries with fragile economies.

National Self-Sufficiency

So, what the industrial revolution and its continuum has done, is to set into place trading systems, and a merchant culture, that it will be difficult to reverse. **The most stable system would be for each nation to provide for itself - to become self-sufficient, only buying from overseas those products which cannot be sourced at home** – but we are a long way from that ever becoming a reality. It is said that the UK at the moment can only produce goods that meet 60% of its needs. Is self-sufficiency a realistic target to aspire to? Could it be reached within the three generations that we have left?

Fig.42
A local farmer's market
(From: clipart)

Britain's Responsibility

As with the Industrial Revolution, Britain is again responsible for setting into play an international trading system that now runs out of control, feeding the IR continuum, and contributing to increasing levels of carbon emissions. Britain started it off but, because it is a small country with limited resources, it has long been left behind by the larger countries with vast resources of mineral and fossil-fuel wealth. Britain tries to keep pace with the larger, resource-rich countries but is really fighting a losing battle. It would be much better placed in leading the world in finding ways of becoming self-sufficient, supporting its own farmers and reducing carbon emissions. And by modifying its economy to support those in most need and in developing green products.

Recently in the news has been the collapse of the UK Steel industry, due to cheap imports from China. Rather than trying to shore up outdated plants, which use fossil fuels to make steel, Britain would be better off using governmental investment to lead the world in developing a carbon-free steel.

Trading and Competition 2

I mentioned earlier in this chapter the competitive spirit that trade engenders. I admit that Britain started trading in this way in the nineteenth century, by making use of its empire links, because it wanted to get a competitive edge over other nations. Other countries, who have followed suit and come to dominate trading systems, have also done so for competitive reasons. Indeed, it is almost impossible to separate the concept of a market economy from the concept of competition and rivalry. But, unless, the nations of the whole world stop competing with their neighbours and reinforcing the IR Continuum, then we will no longer be here to compete against each other.

Global co-operation is what is needed at the moment, not competition; Britain needs to join forces with its neighbours to save the planet.

In a recent TEDx speech, "Why We Need to rethink Capitalism", Paul Tudor Jones II[48], formerly from big business himself, talked about a profit-led emphasis (to the exclusion of all else) that has led to a situation in which the concept of *humanity* has been removed from the corporate world. He said that profit margins, at 12.5%, are currently at a 40-year high and that higher profit margins exacerbate income inequality, with the US having the greatest levels of inequality in the world. He demonstrated a strong link between income inequality and a series of social health metrics. He described a new way of corporate behaviour (The Just Index), in which the public are given a voice.

The Transatlantic Trade and Investment Partnership (TTIP)

TTIP is a series of trade negotiations being carried out mostly in secret between the US and the EU. It is a bi-lateral trade agreement and is about reducing the regulatory barriers to trade for big business and includes things like: food safety law, environmental legislation, banking regulations and the sovereign powers of individual nations. The Independent[49] lists six reasons why we should oppose TTIP:

- The British NHS, as a public institution, is at risk, as one of the aims is to open up Europe's public health, education and water services to US companies, which could mean the privatisation of the NHS;
- Food and Environmental Safety: the TTIP's agenda is to seek to bring European standards on food and the environment, closer to those of the US. But US regulations are much more lenient, with 70% of processed food sold in US supermarket containing ingredients that have been genetically modified. The US also has very lax laws about the use of pesticides and the feeding of growth hormone to cattle;
- Banking Regulations: it is feared that TTIP will remove current restrictions on banks imposed after the 2009 financial crisis;
- Privacy: after a huge public backlash, the European parliament did not agree to an anti-counterfeiting trade agreement (ACTA), which would have allowed internet service providers to monitor people's on-line activities. It is possible that TTIP may bring this back.
- Jobs: the EU has admitted that TTIP may bring in unemployment, as US has weaker labour standards and trades union rights.
- Democracy; this is the greatest threat that would be brought in with TTIP, as it will allow companies to sue governments, if those governments' policies cause a loss of profits.

It would appear that TTIP will allow the big US corporations, already responsible for huge emissions of CO_2, to be given a free reign to wreak havoc in Europe as well.

The Merchant Culture

In the End Piece to my first book and the introduction to this book, I stated that the world had been taken over by merchants – people who trade in all kinds of goods for their own benefit – and how this was destroying the world. I still hold this opinion, 22 years after first making the observation. The world is still controlled by merchants, as well as the greed and acquisitiveness that often accompanies this merchant culture. Unless this is addressed, many of the measures described in this chapter and elsewhere in this book, will make no difference to the domino effect this merchant culture is having on the stability and sustainability of the planet.

A Downturn in Global Trading Systems?

A recent joint publication from the Centre for Economic Policy and Research and The Robert Schuman Centre for Research Studies[50] suggests that there is currently a global trade slow down. The document contains 20 properly scrutinised research papers, which all come to the conclusion that there is a downturn in global trading patterns. Various conclusions are drawn from this; for example, a rise in protectionism, another impending collapse of global markets etc. Economists are obviously worried about this, as they think it will impede economic growth. However, it may herald a worldwide trend in consumers realising there is a climate change crisis and subsequently reducing their consumption of imported goods, deciding not to adhere any more to a throw-away culture.

According to the World Bank, a brief review of the evidence suggests that both cyclical and structural factors have been important in explaining the recent slowdown in global trade[51]. With high-income countries accounting for some 65 percent of global imports, the lingering weakness of their economies five years into the recovery suggests that weak demand is still impacting the recovery in global trade. But they feel that weak demand is not the only reason as trade had become much less responsive to income growth, even prior to the crisis. There is some evidence to suggest that part of the explanation may lie in shifts in the structure of value chains, in particular between China and the United States, with a higher proportion of the value of final goods being added domestically—that is, with less border crossing for intermediate goods. In addition, the post-crisis composition of demand has shifted from capital equipment to less import-intensive spending, such as consumption and government services.

I personally do not think that the downturn in global trade is a disaster; indeed, it may herald a new way forward, which has a glimmer of hope of saving the planet. This whole issue is discussed further in chapters 5 and 7.

SUMMARY OF INTERCONNECTIONS WITH TRADE

CONNECTION	AS SHOWN BY
1. Industrial Revolution	1. International trading systems developed at the same time.
2. IR Continuum	2. As it has continued and the manufacturing base has grown, so has the amount of global trade.
3. Carbon emissions	3. Have increased, along with increased trading across continents.
4. Population increases	4. More people living in the world means more trading and more carbon emissions.
5. Economic systems	5. These are now based on successful international trading and GDP is used to measure a country's competitiveness in trading, leading to an obsession with profit margins.
6. International competitiveness	6. Successful trading forms the basis of assessments.

7. Trading patterns	7. The whole trading system encourages deception by major companies in order to gain a competitive edge, whilst ignoring an impact on climate change (e.g. ExxonMobil, Volkswagen.

CHAPTER 5
World Human Population: Past, Present, and Future

Up until the industrial revolution, the human population in the world had remained fairly static at around one billion people but, by 1930, it had doubled to 2 billion and had reached the third billion by 1959 (in less than 30 years), the fourth billion was reached another 15 years later (1974), and the fifth billion in only another 13 years (1987). According to the most recent United Nations estimates, the human population of the world is expected to reach 8 billion people by the spring of 2024. And 90% of world's population now lives in cities. A big change from the largely agricultural communities that existed before the industrial revolution.

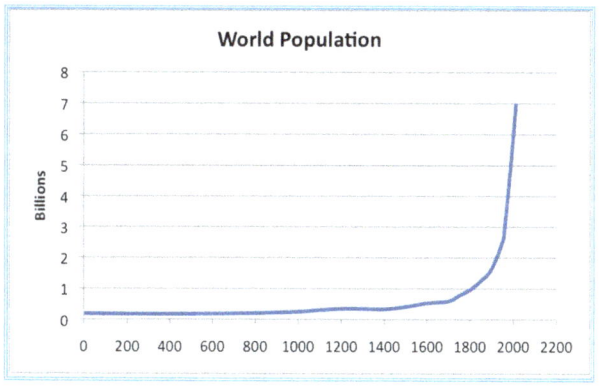

Fig.43 The human population increase since the first century
From: http://www.google.com

Figure 43 shows the human population increase since the first century - an almost identical curve to that in Fig. 7 (Chapter 1), which shows the increase in carbon emissions over the last two centuries. Thus, there would appear to be a very strong link between human population increases and the increase in carbon emissions, perhaps through the common connection they both have with the industrial revolution.

The population of the world is currently growing at a rate of around 1.13% per year, with the average annual population change currently estimated at around 80 million per year. The annual growth <u>rate</u> reached its peak in the late 1960s, when it was at 2% and above. Population statisticians expect the human population to begin to level off at about 11 billion people, which they think will be reached by the end of this century, mainly because family size is reducing. In some Asian countries, for example Bangladesh, family size is now just over 2 children per family, having reduced from about 5.5 children per family 50 years ago; this is mainly as a result of better education about birth control and a demonstrated relationship between large family size, poverty and infant mortality.

The most populous country is China, followed by India, the USA and Indonesia. The pie chart in Figure 44 shows the breakdown by country of those countries with over 100 million people. The United Kingdom ranks 22nd in the world in terms of population size, with just over 62½ million people. The smallest nation listed in Wikipedia is the Pitcairn Islands, with a population of just 48.

As long ago as 1798, Thomas Malthus (1766-1834)[57] warned that population increase might create problems, as his calculations showed that population size increases exponentially (as in the Fig.43 graph), whereas food production increases arithmetically. He thought there would come a time when we would no longer be able to feed ourselves. He predicted that there would be a halt in population increase, followed by a rapid reduction in the population of the world, caused by natural catastrophes, such as famine, disease and war. He made a number of suggestions about what needed to be done to curb this population increase, such as "moral restraint", with criminal punishments for those who had more children than they could support. Some of these ideas led to him losing credibility yet, 200+ years later, we can see that his theory has come to be true, surprising in a way, as he came up with his theory long before the human population began its phase of most rapid increase. And it is interesting that some countries (eg China, Bangladesh) have used his ideas about limiting the number of children they have, though recently China has relaxed this policy because of a shortage of young people to work in their growth industries.

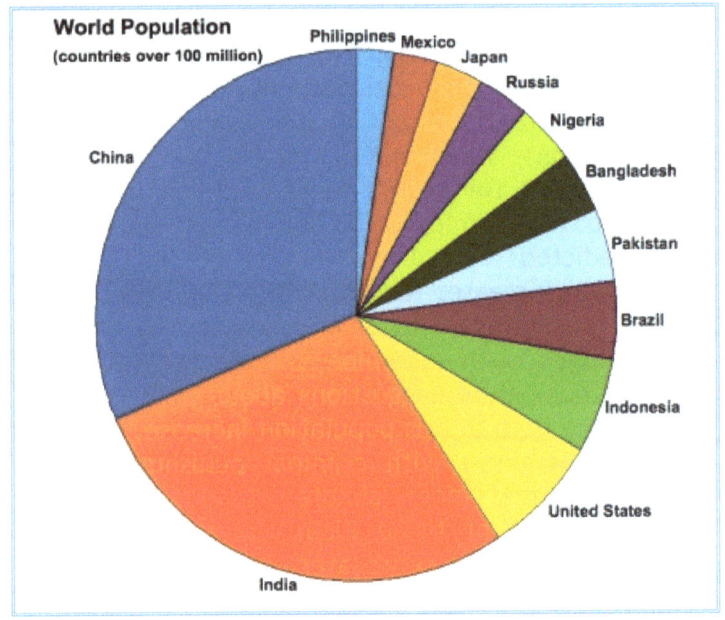

Fig.44
World Population breakdown for countries with over 100 million people from Google Images

A recent article in The Guardian condoning China's relaxation of its one-child policy[53] has met with considerable correspondence, especially by population scientists, who are continuing to urge that it is highly important to do something to curb population increase. However, human rights organisations hold a different view on this issue. And economists urge to maintain the numbers of babies being born, in order to keep future generations to work in the (fossil-fuel-burning) industries to keep the economy growing. Population dynamics can therefore be contentious issue.

Studies of animal populations have shown that population size tends to increase exponentially (as in the Fig.45 graph) until a point is reached when some external factor causes a decline in the population. With some carnivorous species, this shows a regular pattern which is very strongly related to the availability of prey. However, as with food webs, the real situation is rather more complicated than this.

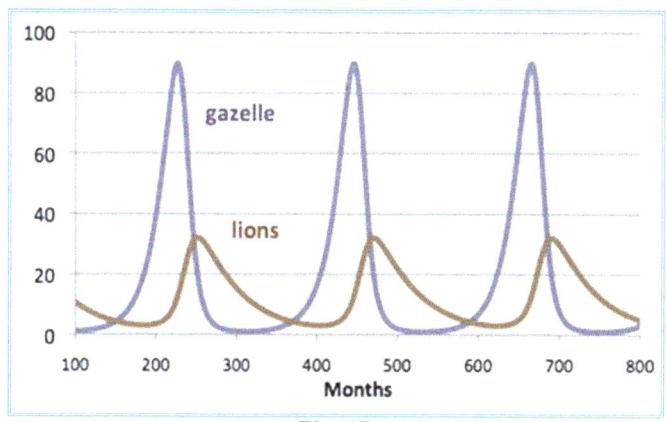

Fig.45
Population size – the relationship between predator (lion) and prey (gazelle)
From:
https://brothersdiamond.wordpress.com/2013/07/31/the-geometry-of-behavior/

In other species, a maximum population size is reached, followed by a slight decline and then a levelling off. During the 1960s, when human population was increasing at its fastest rate, there were great concerns about the future of the human race and whether the earth would be able to produce enough food to sustain all human life. Some recent studies by Professor Hans Rosling[54] and produced in video form ("Don't Panic – the facts about population") have suggested that family size is reducing in most of the world, though not yet in Africa, nor in the UK. His projections show that, by 2050, eighty per cent of the world's population will live in Asia and Africa, as population size is beginning to decline in Europe and the Americas.

The huge increases in population are undoubtedly the reason why some countries are tampering with the genetic code of certain crops, to develop food crops that give better yields, are less affected by pests etc. However, I believe that this is another example of human activity changing the face of the earth that we inhabit, without first checking what the long-term consequencies of this tampering might be.

Population increase is probably related to a number of factors: the discovery of antibiotics, which can have the effect of prolonging human life; vaccination programmes, which have eradicated or reduced the incidence of some of the most lethal diseases (eg smallpox); a higher standard of living since the industrial revolution. There are all kinds of theories promoted by experts in population dynamics as to what point the human population will have reached its zenith and what will be the external factor which triggers a rapid decline in human numbers. It could be anything: overcrowding, leading to wars; climate change, leading to deaths by increased incidences of weather disasters and poor crop yield; a lack of food; air pollution/ lack of oxygen; industrial accidents, especially connected with the nuclear industry; global warming, leading to deaths from heat exposure – some countries (eg India) are already showing record high temperatures and increased numbers of deaths associated with this; the appearance of new "super-bugs", resistant to known antibiotics; a meteor strike; or something else, not imagined as yet. An article in the Open University's "Open Minds" magazine[55], entitled "Humanity's Last Stand" proposed five of the greatest threats to the continued survival of the human race, according to OU experts, as being:

- The appearance of super-bugs resistant to current antibiotics;
- Nuclear war;
- A takeover by robots;
- A hotter planet;
- A meteor strike.

At the moment, we can say that the increase in the world population is one untoward consequence of the industrial revolution and its continuum. We can also say that the increase in the human population has led to increased human activity in those areas which are damaging to our planet. So, there is another interconnectedness here: climate change/species loss connected to human activity, especially since the industrial revolution; increased human population also since the industrial revolution, also leading to species loss as humans take over new habitats. But there are other factors also interconnected with these three factors. One of these is affluence.

Affluence and carbon emissions

Statistics promoted by Prof. Hans Rosling demonstrate a clear relationship between extreme poverty and population dynamics. They also show that, whilst many people are starting to move out of poverty (as a consequence of the better lifestyles of all since the industrial revolution), there are some who cannot manage this without outside help, and these remain in extreme poverty (mainly in Africa and Asia). His statistics show that the richest people in the world have the greatest use of carbon emitting fuels (coal, oil and natural gas), being responsible for 50% of all carbon emissions. Indeed further analysis shows that 85% of carbon emissions come from the medium rich to the very rich. The poorest people on earth, despite their numbers, only contribute towards 15% of carbon emissions. So, this is another interconnectedness – whilst there might still be large numbers of people living in poverty, they cannot be blamed for the carbon emissions causing climate change. In fact, many of them have become the victims of it.

So, we can see that the increase in carbon emissions has been caused by the industrial revolution but that the exponential rise in the human population has resulted in more and more human activity, causing more and more carbon emissions. Because the human population has been rising exponentially, it has set off an accompanying exponential rise in carbon emissions, the rate of which has surprised many scientists. It is only by looking at the interconnections that we can come to these conclusions.

Fig.46
Logo for World Population Day

However, there are many factors at work here. Just as the food chain is rarely a simple chain but more like a food web, so it is with carbon emissions. Increased industrialisation – yes – but also increased numbers of people involved in more and more industrialisation. Because it is more complicated than a simple cause and effect, it will take very sophisticated methods to reverse the trend, which will encroach on every area of life. Also, actions in some countries may not be relevent in other countries, so what is needed is a global response. Some of these issues are discussed by Paul and Ann Ehrlich in their book, "The Dominant Animal; Human Evolution and the Environment" (2008)[56].

Poverty

Bill Gates, founder of the Microsoft corporation, and of the Bill and Melinda Gates Foundation, is an advisor to the Global Povery Project, which campaigns for the elimination of poverty. As part of the Global Citizen Initiative, a number of ambassadors have been appointed, of whom one is my son Ben, to raise awareness of the fact that poverty could be eliminated by 2030, with relatively small increases in aid budgets. GCI challenges some of the myths and beliefs that the world is getting worse and that extreme poverty cannot be solved. It has been said that, for the cost of the Iraq war, we could have ended world hunger for 30 years.

However, the sadness is that, when people do lift themselves out of poverty, they then start adding to the total carbon emissions by buying themselves vehicles and household equipment and gadgets that they previously could not afford. However, I do not think this is a reason not to help people out of poverty. What we must do is educate people about the consequences of human activity and climate change so that they, with the rest of the world, will start to find new ways of living that do not place the planet at risk.

The Super Rich

Prof. Paul Rogers has published a paper for the Oxford Research Group[6,57](Sep. 2012), entitled: "Chances for Peace in the Second Decade – What is going wrong and what we must do." He identifies two root issues which bring about the likelihood of conflict and/or war. They are: Socioeconomic divisions and environmental constraints (climate change), which he considers to be interrelated. In a section headed "Rich-Poor World", he outlines how global economic growth has become more and more unbalanced, leading to the existence of a trans-global elite who own own about 85% of the world's wealth (1.5 billion people out of a world population of 7 billion), with a super-elite of many thousands of multi-millionaires. Because of the size of the elite, it acts as a self-contained global entity and persistently fails to recognise the endemic mal-distribution of world wealth and income. Because of improved education and better communication, many marginalised groups are becoming aware of these divisions and injustices, leading to despair, resentment and anger. All over the world new social movements are developing, to challenge the old order, leading to unrest, conflict and wars.

Another Professor, Andrew Sayer from Lancaster University, has released a book entitled, "Why we can't afford the Rich"[58], a book which won the 2015 Peter Townsend prize. He states in his book, *"We cannot continue to provide the rich and super-rich with unearned income. Their political power is a threat to democracy, and their excessive consumption and dependence on never-ending growth are unsustainable."*

Oxfam has recently released figures[59] that show that, by 2016, the combined wealth of the richest 1% of people will have overtaken that of the remaining 99% of people. One in nine people in the world do not have enough to eat and more than a billion live on less than $1.25 a day.

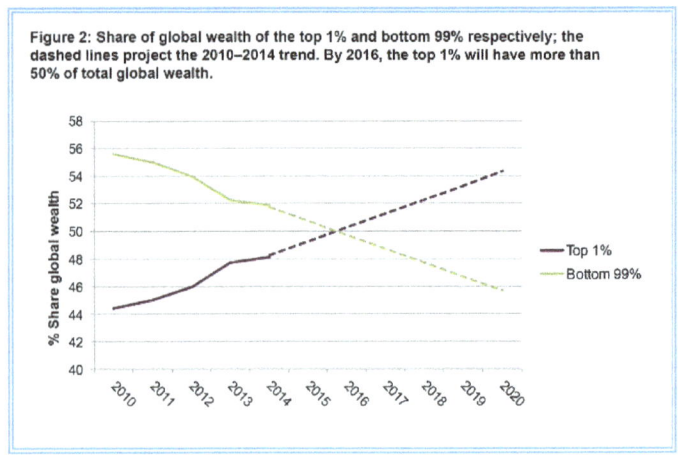

Fig.47
From: graph taken from statistics provided by Oxfam
(http://www.oxfam.org.uk/blogs/2015/01/richest-1-per-cent-will-own-more-than-all-the-rest-by-2016)

Fig.48
From: lukeflegg.wordpress.com

There are those who are of the belief that big business is entirely responsible for terrorist attacks.

In my first book I argue that it was no coincidence that the 9/11 outrage was targeted at the World Trade Centre – an icon for big business. In her documentary, "The Economics of Happiness", Helena Norberg-Hodge[60] sets out how the world is moving simultaneously in two directions: Government and big business continue to promote globalisation and the consolidation of corporate power, whilst the rest of the world are resisting those policies and working to forge a very different future, which involves re-building on a more human scale, with the localisation of economics being the goal.

Urbanisation

Concomitant with the increase in the world's population, there has been an increasing trend in people moving to cities, in search of work, so that now 90% of the world's peoples live in cities. A good example of this trend occurred with my own family ancestors on my father's side, who were originally farm labourers in Norfolk. During the mid-1800s, due to imports of vast quantities of American wheat, many British farms went out of business and my ancestors found themselves out of work. My great great grandfather, John Jackson, and his large family, travelled north to Lincolnshire and then Yorkshire, finding work eventually in the coal mines and coke industry, thus contributing unwittingly to the whole industrialisation process. And a similar story can be told over much of the world, as factories have replaced farms and the IR Continuum spreads.

Industrialisation of Farming

But, there have also been trends to make farms become more like factories, with the introduction of battery farms for the raising of poultry and egg production, the keeping of pigs in restricted metal cages, whilst they give birth, and the transport of live animals across continents under inhumane conditions and without water, only to be slaughtered abroad in sometimes brutal practices. There has been limited success in abolishing some of the most brutal of practices, by campaigning groups, but there is still a long way to go before animals are seen as related species sharing this world with us, rather than commodities to be sold and slaughtered for profit.

Farmers have also utilised more of their land, with the pressure to become more productive, so that hedgerows have disappeared and, with them, many of the bird species that nest there, along with small mammals[61]). However, legislation has been introduced to control this[61].

Some of the inhumane consequencies of the industrialisation of farming

Fig.49
Battery hens

Fig. 50
A sow being kept in a restrictive pen after giving birth

Fig.51
Sheep being shipped by ocean in very cramped inhumane conditions

Another of the situations predicted by Malthus was the development of "gluts" when farm production over-supplied with certain crops and foodstuffs. In Europe there have been gluts of sugar, butter, wheat and now milk. And at the time that I write, UK dairy farmers and going out of business because supermarkets are failing to pay a fair price for milk, because they can buy it more cheaply elsewhere. The Fair Trade system was introduced a few years' back, to help developing countries sell their goods at a fair price; this sytem also needs to be introduced for dairy farmers in the UK.

In some countries, people have stayed in rural areas and continued to till the land, but because of increased populations, there is less land available. This has resulted in the felling of forests, in order to produce agricultural land. And this itself has affected the climate, as these very same forests were the major "sinks" in which carbon dioxide in the atmosphere was absorbed to produce oxygen, through the process of photosynthesis (chapter 1). The loss of the forests has also led to the loss of many rare species , as their habitat is lost (chapter 1).

In some countries too, there has been a move to fell forests to grow illicit crops, which are more lucrative, thus fuelling the drugs cartels and parallel economies.

Values

One of the factors associated with a heterogenous world population is that people in different parts of the world grow up with different values and will seek to adhere to those values, wherever they live. It is part of their cultural identity. Thus, many in the UK adhere to a *protestant work ethic*, which may have been behind the industrial revolution. Their value system also has much in common with Christian values of justice and fairness, even though many people no longer have a faith. In this country we also value education. This is, of course a generalisation, as many from the business world born out of the IR Continuum, do not adhere to the values of justice and fairness but instead see the acquisition of money as their main goal, perhaps because of ancestral links with the landowners of our feudal past. They are not alone in this, as several of the other cultures settling in this country have different goals and values, mostly based on family and/or the acquisition of wealth. This may be because they have come from poverty in their own countries, and want to send money to their relatives back home, or because they have come from a value system which admires those who have made a lot of money, rather than those who are well educated. Those coming from cultures which value the family (and father) above all else may find it difficult to adjust to a system based on the employer/employee relationship. In a multi-cultural society like ours, this can lead to conflicts, disaffection and a lack of awareness of the grave issues facing the world today.

Fig.52
Multi-cultural Britain

The motivation to make money is a strong one and can also lead to a denial of the seriousness of climate change.

Many young people born in this country of parents who emigrated here from commonwealth countries, find themselves torn between two cultures: the culture which their parents still value and the British culture. They may feel they do not belong to either and this can lead to disaffection and the attractiveness of joining terrorist groups or criminal gangs, to which they feel they can belong. It is a development with which we have not yet come to grips in the UK – and a similar situation exists in France, Belgium and Germany and other European countries. I worked in an inner city area of Birmingham for some years, trying to help the unemployed find work, most of them young, ethnic minority men. It was very difficult as most potential employers just did not want them. I saw many becoming disaffected and angry and others turning to drugs and/or crime. I do not find it surprising therefore that some young people have been radicalised and have joined terrorist groups.

I believe that it is differences in our value systems, way of life, mother tongue, clothing and appearance that lead to disharmoney between ethnic groups and can, if we allow it, lead to racism.

Any strategy for the future needs to acknowledge that, whilst we all have in common our humanity, we may adhere to different value systems – and these need to be respected, if we are to move to greater co-operation and joint efforts to save the planet.

Migration issues

Some of the experts who have tried to predict the future of the human race, as a consequence of the multiple effects of population explosion, climate change and war, have foreseen that, in the future, there would be massive migration from Asian and African countries to Europe, Australasia and America. As I write this, it would appear that this migration has already started, with thousands of refugees fleeing from the Middle East, Asia and Africa to Europe, at a rate with which the European countries cannot cope. If the experts are right, this is only going to get worse, with more deaths from drowning in the Mediterrean Sea and more exploitation from ruthless traffickers who are keen to make money out of this human crisis. It has already led to conflict and division between different European countries about how best to cope with the situation. Some want to close their borders completely but this strategy is just as inhumane as the practices described earlier in treating farm animals as if they were factory commodities. The wars in Syria, Iraq, Sudan, Somalia and Afghanistan have resulted in huge numbers of refugees, including pregnant women, the elderly and children, fleeing the violence in order to find a safe place to live and trekking hundreds of miles across countries to reach their preferred destination.

According to George Monbiot[62], one of the likely catalysts for the 2011 uprising in Syria was a massive drought – the worst in the region in the instrumental record – that lasted from 2006 to 2010. It caused the emigration of one and a half million rural workers into Syrian cities, and generated furious resentment when Bashar al-Assad's government failed to respond effectively. Climate models suggest that man-made global warming more than doubled the likelihood of a drought of this magnitude.

Wars in North African countries have also led to large numbers of people fleeing their countries, trying to cross the Mediterranean in flimsy boats, to get to Europe. Thousands have drowned. Others have ended up in the Greek islands and Greece, already in dire economic circumstances, has had to cope with helping the refugees as best they can. Yet others, trying to get to Britain across the English Channel, are trapped in makeshift camps in Calais.

Fig.53
Refugees trying to reach the safety of other countries

Communication

The computer age, and especially the development of the internet, has led to significant changes and advances in communication. People are now in touch on a regular basis through social networks and email. This has transformed the world, both for good and for bad. Whilst the media still try to control the news and impose their own biases on the public, and have a modicum of success in this, people are also receiving information informally through other networks. Thus, demonstrations can be organised very quickly and, in some instances, this has brought down governments, as in the Arab Spring and in Ukraine. There is currently a world-wide disaffection with politicians, who are seen as corrupt and not trusted any more. Unfortunately, bringing down a government, or a despotic leader, does not always lead to the changes people are seeking, as others come in to fill the vacuum, that are also unpopular and/or corrupt. There is thus, at the moment, considerable unrest and instability throughout the world and it is difficult to predict where this will end.

It is also difficult to predict where the new-found ease of communication through the internet will take us. Let's hope that, ultimately it will be for the betterment of this planet, its peoples and its wildlife.

Overcrowding

In some, well-publicised experiments with rats in the 1950's, Calhoun found that when rats were kept in extremely crowded conditions, but with unlimited water and food and protection from predators, there were a number of changes in their behaviour. Male rats in the most crowded pens became violent and aggressive, "going berserk, attacking females, juveniles and less-active males." There was also "sexual deviance." The mortality rate among females was extremely high and there was a breakdown in maternal behaviour. Mothers stopped caring for their young, stopped building nests and even began to attack their offspring, resulting in a 96 percent mortality rate in the most crowded pens. Parallels were drawn between these experiments with rats and whether the same could be said to be true for humans, in particular those who lived in cities (Calhoun, in the Scientific American 1962). There were a number of reviews and other experiments following this, which concluded that it was too much social interaction that caused the pathological behaviour, rather than the overcrowding. The studies have been reviewed Carla Garnett[63].

A Civilised Society

At the beginning of this book, I included a list of factors important in a civilised society, put together at the turn of this century by Barbara Panvel and me (Table 2). I believe that any developments that are made in the future to rescue the planet are carried out with this list in mind. Indeed, these characteristics may become even more important as we, as a global population, seek to find ways to co-operate more closely to save the planet.

Much of the unrest caused by better communication through the internet is because the main concerns of ordinary people are about corruption in leaders and unhealthy alliances between politicians and big business. Some of the issues of importance to ordinary people are just not being taken on board by politicians and leaders, sometimes to their cost. Richard Douthwaite, in his book "*The Growth Illusion*" (1999)[5,64] provides data from research that shows that most people when asked about what is important in the quality of life that they lead, come up with issues that are, in the main, not related to how much cash is available to them. Things like:

- The quantity of goods and services produced and consumed;
- The quality of the environment they live in;
- The fraction of their time available for leisure;
- How fairly (or unfairly) available income is distributed;
- How good or bad working conditions are;
- How easy it is to get a job;
- The safety of their future;
- How healthy they are;
- The level of cultural activity, the standard of education and ease of acces to it;
- The quality of the housing available;
- The chance to develop a satisfactory religious or spiritual life;
- The strength of one's family, home and community ties.

There is much in common between this list (from Douthwaite) and the list that Barbara and I produced (Table 2) of the characteristics of a civilised society. Also, a recent report published by the New Economics Foundation[65] has used a shorter list to determine the UK's success in economic terms, under the headings: good jobs, wellbeing, environment, fairness and health. A summary from that report and an extensive quote from Douthwaite are included with Chapter 7, on the economy.

Certainly in the UK, there seems to be an obsession among politicians about the economy and growth but little concern for issues in Douthwaite's list, nor the effects of global warming and climate change, nor for people in poverty, nor for the many refugees fleeing their homes because of warfare there. The anomaly is that, for the UK at least, some of these wars people are fleeing from were caused by us messing in those very countries from which they are fleeing and actually making things worse for them. People fleeing from war-stricken countries do so in the hope that some of the things in the three lists may be available to them elsewhere.

Population increase and the future of the planet

At the beginning of this chapter, I mentioned that population scientists believe that the human population of the world will level off at 11 billion people at the turn of the century. Unfortunately, this is about the same time that climate scientists are saying we may be facing a mass extinction of species (three generations into the future), which will have a significant effect on the human population. If we are going to do anything about all of these interrelated issues, it needs to be now – it will not wait until three generations' time. Something needs to be done to limit or reduce population increase.

INTERCONNECTIONS WITH POPULATION DISCUSSED IN THIS CHAPTER

CONNECTION	AS SHOWN BY
1. Industrial Revolution and its Continuum	1. Population increase
2. Increased carbon emissions	2. Population increase
3. Climate change	3. More people engaged in more activity, which results in more carbon emissions; climate change causing droughts and extreme weather events, resulting in large numbers of migrants moving to cities and/or other countries.
4. Urbanisation	4. 90% of global population now living in cities, all contributing to activities which increase carbon emissions
5. Poverty	5. Climate change affecting agrarian communities in some parts of the world; market economy favouring the rich at the expense of the poor; a global elite owning 85% of the world's wealth.

6.	Affluence	6.	85% of carbon emissions come from medium rich to very rich.
7.	Internet	7.	Better communication leading to disaffection with leaders and politicians, civil wards, refugees and migration to other continents.
8.	Industrialisation of farming	8.	Loss of hedgerows and thus species; animals kept inhumanely; animals exported across continents in inhumane conditions.
9.	Population increase	9.	Greater use of antibiotics and vaccination, leading to reduced mortality; more people engaged in trade and activities which increase carbon emissions.

CHAPTER 6
Conflict, conquest, weaponry, wars and the power of propaganda

Ever since the Stone Age, man has made himself weapons, initially to help slaughter animals for food but then to fight other tribes to gain power one over the other. Because of human ingenuity, over the centuries more and more sophisticated weapons have been developed, in order to gain supremacy in conflicts between individuals and groups. So, what may have started with hand-held stones and clubs, escalated into bigger and bigger, more sophisticated and more powerful, weapons, until we are where we are today, with nuclear weapons that can destroy humanity, chemical and biological weapons of mass destruction, that can do the same, and missiles that can be fired from distant places, well away from the battlefield, and guided remotely to their target.

The escalation and production of weapons was facilitated and enhanced by the industrial revolution until, towards the end of the Second World War, the nuclear bomb was developed – and released over Japan – which has the potential to destroy the population of the whole world, human, animal and plant. The recognition of the potential of this lethal weapon to trigger a nuclear war, which would destroy us all, led to the Cold War and a tacit agreement not to use nuclear weapons. But this has not stopped conflict and hatred between nations, suspicion, a lack of trust between nations and the desire to find a (safer) weapon to demonstrate one nation's perceived supremacy over another. The latest development is the use of satellite-guided drones and missiles, to wipe out targets (people) from a distance, but many mistakes have been made with these, with innocent people and children being slaughtered unnecessarily.

The purpose of this book is to show the interconnectedness of all things, not to get into an argument about the rights and wrongs of wars, but I think it might be worth a look at some of the psychology behind the use of weapons, if we are to find a way of moving to global co-operation in order to save the planet.

I believe that the Hollywood film industry has a lot to answer for in its portrayal of "macho" men, firing guns in order to wipe out the "enemy", both in their early Westerns and, more recently, in promoting different kinds of aggression, associated with maleness, as being the norm. Many young men watching such films are influenced by this and identify with the macho culture, so that it becomes part of their sexual identity, much the same as the glamour culture clearly influences young women.

The parallels that can be seen between a potent form of male sexuality and the firing of a gun are obvious to all, though rarely acknowledged. Examples of this are the rapidly-increasing numbers of mass killings in the USA by young post-pubertal men wielding repeating rifles and the unwillingness of the American male gun lobby to support legislation banning guns from general use, as it is seen as a form of emasculation.

This macho culture, rivalry and the desire for conquest has even crept into sport, with F1 and other victory celebrations often using a bottle of champagne, shaken to mimic ejaculation.

The Hollywood macho film culture has been exported successfully overseas too, with young males in many countries identifying with this aggressive gun culture and the need to conquer and eliminate rivals.

I am, of course, seeing this from a woman's point of view and acknowledge that it is not only men who are responsible for instigating, or participating in, wars. Britain's Margaret Thatcher set in motion the Falklands War, which led to thousands of needless casualties from Argentina and the UK, when agreement could have been reached through diplomacy.

A hero image is frequently propagated to entice young men to fight for their country, even though the reality of war is very different from the glamorised image. During the First World War, many young men (some barely out of childhood) volunteered to go to war, only to find that trench warfare was far from glamorous and many of them, if they did return home, never really recovered from shell shock (or post-traumatic stress disorder). Yet, those who opposed the concept of war at that time, were frequently put into prison and scorned as cowards. Those who deserted the army were often executed.

So, why is this hero/glamour myth taken on board by so many? Why does the glamour/hero culture perpetuate despite the shocking experiences of the First World War and later wars? I think the rise of the Hollywood gun culture during this period has a part to play here. Plus a mind-set in politicians that a successful war will enhance their reputation, each aspiring to be a modern-day Churchill or Nelson.

Fig.55
A young soldier firing an automatic rifle during the Afghanistan war

The arms industry

Another factor at work here is the arms industry. Like other businesses, which have international trading opportunities, there is an assumption that selling arms to other countries, even to those who are potential enemies, will be beneficial to our country, to the business, jobs and to the economy. It is a contradiction and leads to the development of more and more sophisticated weaponry, all leading to more carbon emissions, greater conflict and more mass killings. The economic reasons, used as an excuse to allow the escalation of the arms industry, are about encouraging economic growth will be dealt with in the next chapter.

Power and prestige

There are power issues at work here too. For example, the USA sees itself as a super-power and certain right wing elements within that country desire to see this power maintained or enhanced. For example, at the end of the Cold War, when the

USSR was split into individual countries, some power-hungry American groups thought that this would be an ideal time to increase their power in the world. There have been suggestions that, even before the Iraq War, there was a consensus among these right-wing groups that Iraq should be a target, with regime-change and the elimination of Saddam Hussein being a priority. The war, of course, was a disaster, especially for the people of Iraq, with a vacuum being left there, which has been filled by anti-west terrorists, causing thousands of refugees to flee to Europe for their safety. There are also historical power issues associated with Russia and its control of former-USSR countries, as shown by the invasion of the Crimean region of Ukraine in 2014.

Propaganda

And suspicions between countries, who were once at war, linger on and are exploited by a xenophobic irresponsible media hype. Propaganda has been used to increase suspicion, racism and xenophobia between nations. It was used very much in the two world wars to encourage and recruit young men to join the armed forces ("Your country Needs You" etc.) and used during the wars to wrong-foot the enemy. But today, it is frequently used by the right-wing media to drive and panic our population into open hostility towards people of other nationalities and ethnic groups. Most of what is written is lies but, if repeated often enough, it is believed and taken on board by the gullible, precipitating a fear that we are being swamped by foreigners, who are taking our jobs and our housing. It has led, in this country, to the formation and popularity, of the UK Independence Party. People with racist tendencies have flocked to it in huge numbers.

And some in Britain still hanker for a return to the power we formerly wielded over the British Empire. Questions have been asked about why Britain still needs nuclear weapons and whether it is moral to commit some £80 billion on the Trident programme at a time when benefits are being cut for the disabled and others living in poverty, in the name of 'austerity'. The defence of the country is usually the explanation for this significant outlay but others believe that it is not about defence at all but about prestige. Trident (together with worldwide military capacity) is a badge of power and, without the Trident submarine and its ability to launch nuclear warheads, our presence on the UN Security Council and other international bodies might be questioned. Successive governments have been only too willing to use taxes to promote and maintain this profitable and prestigious industry.

Fig.56
From: http://www.google.co.uk

A few months ago, there was an interesting report in The Independent[66] that a senior serving general in the British army had warned that the government could face mutiny from the army if it were to downgrade its weapons. The unnamed general was quoted as saying that, if the government tried to scrap Trident, pull out of NATO or announce plans to "emasculate or shrink the size of the armed forces", they would be challenged. An interesting use of semantics in view of my earlier comments about the association between male sexuality and weaponry. Needless to say, the Ministry of Defence stepped in to say that we live in a democracy, so the general's implied threat of a *coup d'état* was unlikely to take place. The incident does, however, further reinforce the theory of a strong link between male sexuality and the desire to possess and use powerful weapons.

Friendship between nations

It is these kinds of attitudes which prevent moves away from xenophobic hostility to international co-operation. However, the Queen's example of developing friendship between commonwealth countries could be seen as a model for moving away from domination and control to friendly egalitarian partnerships.

Nelson Mandela's example of relinquishing the desire for revenge on his release from many years' of imprisonment, is another example of what can be achieved. His legacy of using peaceful means to achieve greater racial harmony in South Africa has had a far-reaching (and global) impact.

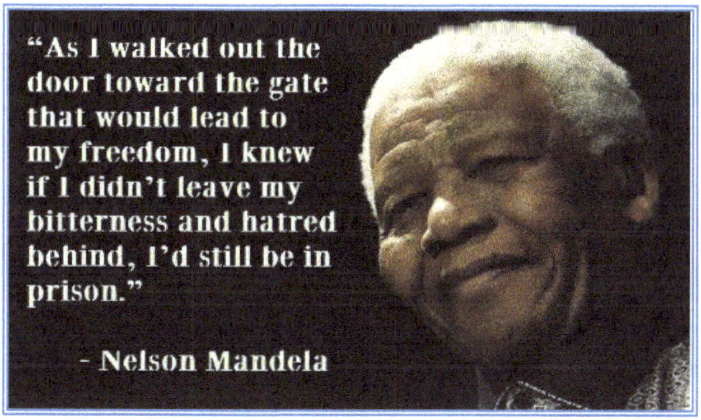

Fig.57

From: http://www.google.co.uk

Any military strategist will tell you that if, during a battle, the circumstances and priorities change, then a new strategy must be quickly developed. I believe that we are now in such a situation in the world, with climate change and the imminent destruction of the planet completely changing the goal posts. It is now totally irrelevant to be maintaining nuclear weapons (of mass destruction), whether for prestige or for defence, at a time when we should be rallying together globally to avert an impending global disaster.

The carbon footprint of war

It has been estimated that the carbon footprint of a small nuclear exchange in a nuclear war would release 690 million tonnes of carbon dioxide, through the burning of cities; this would be more than the current annual emissions of the UK.

But a war doesn't have to be nuclear to have a large carbon footprint. It seems obvious to me that the detonation of large bombs and the fires that are frequently caused in the aftermath of an explosion, are adding substantially to the overall carbon emissions. It has been suggested[67] that the US military operation in Iraq may have clocked up around 160-500 million tonnes of CO_2e, plus a further 80 million tonnes for the healthcare of injured troops. If the coalition forces' activities are added to this, and the effects of a poorly resourced insurgency, this might increase to 250-600 million tonnes. And that's excluding the direct emissions from the explosions that took place.

If we are to find a way out of the global, climate-change, crisis that is affecting people of all nations at this time, we have to find a way of moving on from old hostilities, rivalry and mistrust to a new form of global co-operation. Other great, international icons have also sounded a similar note, suggesting peace instead of warfare as the superior route:

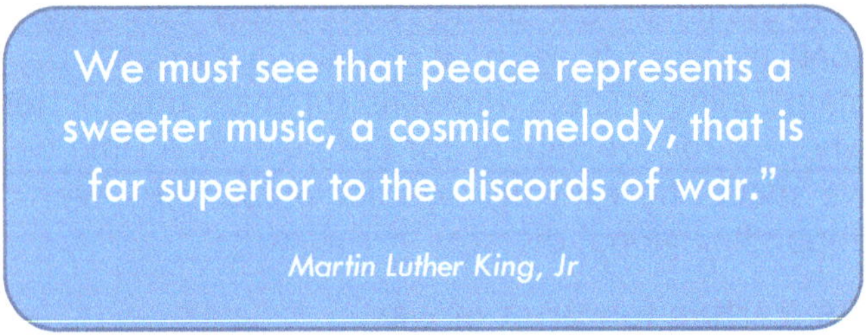

We must see that peace represents a sweeter music, a cosmic melody, that is far superior to the discords of war."

Martin Luther King, Jr

Fig.58
From: http://www.google.co.uk

And Gandhi said, "There is no path to peace. Peace **is** the path."

Surprising in a way that <u>none</u> of these three great iconic statesmen of the past, who promoted peaceful methods of achieving societal harmony, came from the white communities. Is that a lesson for us? Is the white community to blame for the aggressive lack of harmony in the world?

The relationship between poverty and war

Professor Paul Rogers, in his briefing paper to the Oxford research Group (2012)[6,68], argues that, in recent years, there has been significant economic growth across the world, with a super-elite of very rich people developing alongside (but totally unaware of) large numbers of marginalised and exploited people, who have not shared the fruits of economic success. The super-rich have become so through trading systems dominated by transnational corporations, which produce low-cost commodities at the expense of poor farmers and miners across the world. Because of improved communications, a consequence of this is that poor people are becoming more aware of their own marginalisation and have developed new social movements. There has been widespread anger and frustration particularly manifesting against the Middle East autocracies, leading to the so-called Arab Spring. Rogers expects this to grow across the world, and to combine with a growing environmental awareness of the damage that is being done to global ecosystems by powerful corporations. He believes that the socio-economic divisions alone, even without the environmental constraints already manifesting, point to a very disturbed future, with a greater risk of revolts from the margins, leading to wars.

It has also been calculated that, for the cost of the Iraq war, we could have ended world hunger for 30 years (John Greenberg, 2014)[69]. It would seem that those in power have got their priorities all wrong.

So, in summary then: the production of modern weaponry in itself contributes no more than other products of the IR continuum to increasing carbon emissions, but their use is strongly likely to do so. What war does is to promote a cycle of defensive aggressiveness and hostility, which adds to the escalation of more and more sophisticated weapons systems and increasing suspicion and division between nations. Unless this cycle is transformed, redirected or even better reversed, then the global co-operation needed to prevent the destruction of the planet will never happen.

SUMMARY OF INTERCONNECTIONS WITH WEAPONRY AND WAR DISCUSSED IN THIS CHAPTER

CONNECTION	AS SHOWN BY
1. The drive to produce more and more powerful and sophisticated weapons	1. Enhanced by skills acquired during the industrial revolution and its continuum.
2. The use of weapons and weaponry in war	2. Increased carbon emissions.
3. Link between the male sexual drive and the development and use of weapons	3. Disarmament seen as emasculation.
4. Right-wing media propaganda induces fear, rivalry and divisions and hostility towards people of other races and nations	4. Desire and pressure to maintain powerful weapons, including nuclear weapons.

5.	Wars caused by politicians wishing to show their power and to acquire oil resources from other countries	5.	Increased carbon emissions.
6.	Transnational trading systems dominated by multi-national corporations and the super-rich	6.	Marginalised communities starting to take action against oppressive regimes and the super-rich.
7.	Export of arms abroad	7.	Links to trading systems and the economy.

CHAPTER 7
The economy

The economy sometimes seems like a mysterious thing to ordinary people – something that is hard to understand – but there is nothing mysterious about it really. The Oxford English Dictionary defines it as "the state of a country in terms of the production and consumption of goods and services and the supply of money". So, in everyday language it is how *we, as a country, make use of available money, to ensure that everybody has enough to live on.* In a way, it is about balancing the books on a national scale. It is about money.

Fig. 59

I have been able to discover **five different types of economy**. Before the industrial revolution, local communities in the UK lived largely in farming cultures and the economic system was structured around this. But this changed significantly as the industrial revolution gained momentum.   An

144

agrarian economic system has also been called *A Traditional Economy,* and some countries which did not become industrialised, still use this type of economic system. Other countries, which followed Britain in becoming industrialised, developed economies based on trading links and, like the UK, developed *A Market Economy,* which is largely regulated by demand and supply. For some, a market economy is another way of describing *Capitalism.* Some countries have a mixture of traditional and market economies, called *A Mixed Economy.* Yet other parts of the world have government control of their economies and this has been termed *A Command Economy* or *totalitarianism*; this would include countries like North Korea.

A recent article by Pat Conaty describes a *Collaborative Economy for the Common Good*[70]. He suggests that co-operatives and social enterprises are bringing a new dimension to national economies and have been more successful in delivering growth than market economies (in Germany, Italy, Scandinavia, Canada etc.). Some call it community economic development whilst others call it a social economy. From this has developed a *Solidarity Economy,* which seeks to secure systemic change by organising small business and self-employed networks, bringing them into a collaborative economy movement. Such an economy is gaining strength across Europe, as it is based on serving the welfare of people and planet.

The relationship between trade and economies

In chapter 4, I discussed trading systems and how market economies first developed. One suggestion mooted there was that each country, instead of getting involved in complex and comprehensive international trading, should seek to be self-sufficient, only importing goods that they cannot produce themselves. However, I am pragmatic enough to realise that this is not going to happen overnight, as there are too many vested interests in the business world. So in this chapter, I will be looking at other measures that could be introduced, by modifying a market economy to a different (and new) form of economy, which helps to reduce the escalation of carbon emissions.

An uncontrolled market economy

But first, I want to look at the reality of what happens in an uncontrolled market economy. One way of doing this is to look at the measures introduced in the UK by the Conservative government since 2010, which wholeheartedly supports a market economy. These measures are set within the context of a belief that encouraging big business is the only means of making a country wealthy. Thus, the government introduced measures that enhanced the opportunities of the business world to make money: cutting taxes for the richest 4% and for corporations, whilst reducing the amount spent on benefits for the disabled and for the poorest in society. The measures were introduced in the name of so-called austerity which, the government argued, is necessary for reducing the deficit in the balance of payments. In practical terms, the outcome of this is that the government is giving away to the richest people in Britain more money than they are saving by reducing benefits for the disabled and the poorest in society. This does not make sense in a civilised society, as it will lead to further divisions and discontent in society, with the poorest becoming ever more militant.

Fig.60   used by permission

And big business, empowered by the extra money they have been given, use it to continue manufacturing and selling those products which add to emissions of carbon products and other pollutants. This enhances the rate at which climate change is escalating. So, in addition to being in the ever-speeding Industrial Revolution (IR) Continuum, our present government is encouraging businesses to feed that continuum, so that it runs ever more quickly out of control, producing ever more carbon emissions in the process.

A sensible economy in today's circumstances needs to reduce carbon emissions, encourage businesses which produce goods and services which reduce carbon emissions and maintain its

benefits systems for the poorest and most disadvantaged.

A market economy provides unfettered freedom for businesses to carry out their activities, with little government control and little expectation that they will show responsibility for those less well off than themselves, or any responsibility for restricting climate change. And the UK is not alone in encouraging this. It happens in most of the industrialised countries of the world, which are in vigorous competition with each other. Businesses like this freedom of course, to make as much money as they can, but this should not be at the expense of the planet, nor of the poorest in our society.

David Cameron swept to power in the UK in 2010, saying that his would be the "greenest government ever". Jonathon Porritt, in his article: "The Coalition Government 2010-2015; The Greenest Government Ever: By no stretch of the imagination"[61] has demonstrated that, in fact, carbon emissions increased during that term of office (2010-15). This has been reinforced by an article by Michael Le Page in the New Scientist, entitled "Ungreen and not-so-pleasant land"[62]. Le Page provides statistics that show that the UK is **not** on track to meet its climate goals (agreed in Kyoto Summit 1997) and that, rather than increasing its efforts to do this, the government has blocked a series of green measures, thus leading the country even further off course. Perhaps the most contentious is the proposed axing of feed-in-tariffs (FITs), which were available to people investing in solar panels for their domestic electricity needs. Because of these changes the UK has now slipped from 8th to 11th in the RECAI table[73]. (Renewable energy country attractiveness, published by www.ey.com), as shown in Table 4. These RECAI tables change over time but it is not clear why the countries listed on page 54 do not appear on the RECAI list.

Rewarding the rich is not the only way of creating wealth for a country. A recent article by Donald Braben, also in The New Scientist[74], stated that it can be demonstrated that <u>innovation</u> is more likely to produce growth than existing market economy methods. His thesis is based on the history of scientific discoveries which, indeed, started off the industrial revolution in the first place. He has shown that some of the biggest scientific discoveries in our history led to the greatest growth in the economy. If this is true then, rather than funding big business, we should be funding research into new innovatory discoveries, such as carbon-free steel. I would add a rider to this, that the innovation encouraged in this way should also be about reducing our reliance on fossil fuels.

People who like to make money, in any part of the world, appear to have a mind-set that it is their right to do so without hindrance. Many have no conscience about the impact of their money-making on others and have little compassion for those who are the victims of their acquisitiveness, whether they are those in poverty, members of the animal kingdom or, indeed, the whole planet (see also in Chapter 4 – Paul Tudor Jones II).

TABLE 4

RECAI List of renewable Energy Country Attractiveness (first 15 countries in the list) as at Sept 2015

(http://www.ey.com/Publication/vwLUAssets/RECAI-45-September-15-LR/$FILE/RECAI_45_Sept_15_LR.pdf)

Rank	Previous rank	Market	RECAI score	Onshore wind	Offshore wind	Solar PV	Solar CSP	Biomass	Geo-Thermal	Hydro	Marine
1	(2)	US	75.0	2	8	1	1	2	2	3	8
2	(1)	China	74.2	1	2	2	6	1	12	1	16
3	(4)	India	65.9	3	16	3	5	15	14	9	11
4	(3)	Germany	65.7	6	4	6	27*	8	13	15	27
5	(5)	Japan	63.2	18	9	5	26	4	4	4	10
6	(6)	Canada	60.8	4	13	11	23	13	18	5	6
7	(7)	France	59.9	7	7	9	27*	9	15	12	4
8	(9)	Brazil	58.2	5	25	8	8	3	32	2	24
9	(11)	Chile	56.3	19	22	4	2	21	8	17	14
10	(12)	Netherlands	55.1	11	3	26	27*	10	24	32	30
11	(8)	UK	55.0	13	1	16	27*	7	20	25	5
12	(13)	South Africa	54.9	15	28	7	3	33	35*	18	19
13	(10)	Australia	54.2	23	19	10	12	20	10	24	12
14	(14)	Belgium	53.4	26	6	20	27*	11	21	29	31*
15	(17)	Turkey	53.1	9	24	24	11	34	6	6	20

*Joint Ranking

Tax evasion (mainly by the rich) is a major source of lost funds for the economy (£5 billion a year), yet it is often the people who need to claim benefits who are castigated for playing the system. A recent article by James Bloodworth in The Independent[75] showed that **four times more money is lost to the economy by tax evasion than by benefit fraud,** though the difference may be even greater than this if incompetence within the DWP is taken into consideration. He described this as a double standard, with one rule for the rich and another for the poor.

Fig.61
with permission from David Baldinger

In the last few years in Britain a number of scandals have been uncovered, all related to this same acquisitiveness: MPs claiming expenses they were not eligible for, bankers and chief executives getting bonuses, even when they have failed in the job; companies and individuals avoiding the payment of taxes, by using offshore business accounts – and so on. All of these are linked to the same acquisitiveness that fuels a market economy. And the general public in this country have had enough of this. They want to see some honesty – in politicians and in big business – and to see signs of the responsibility referred to above.

A recent scandal has been leaked about the super-rich hiding away their fortunes in tax havens, with details of the names of some of the people who are doing this. As a result, the Prime Minister of Iceland has had to resign and this may be followed by other resignations. Yet, we knew 6 years ago, in a study reported by Heather Stewart in the Guardian[76], that a staggering 21 trillion dollars has been lost to the global economy through tax revenues, as it has been stashed away in tax havens.

Fig.62
From:
https://watershed2015.files.wordpress.com/2015/09/broken-britain-3-mps-bankers.jpeg

Yet it is encouraging that, recently, the Governor of the Bank of England, Mark Carney[77], made some warning statements in a dinner speech, that the impact climate change could trigger a new financial crisis and derail the economy, as it currently stands. He based this analysis on the effects climate change has had in Britain on the insurance industry. Since the 1980s, the number of weather-related events, such as storms and floods, has tripled and the cost to insurers has increased from £6.5 billion to £33 billion, mainly to cover the cost of damaged property and of disrupted trade. He warned that, once climate change becomes a defining issue for financial stability, it may already be too late.

What he identified is that much of the current financial investment is in companies which will be affected by climate change, such as energy suppliers, insurance companies, and oil suppliers etc., whose share prices will fall as climate change begins to bite. If these companies then fail, due to the pressures on them, the value of all kinds of investments, such as pensions and savings, could be affected. In the case of energy companies, if they do not convert to sustainable forms of energy, the pressure to reduce carbon emissions will also make them vulnerable to a reduction in the value of their shares. Oil companies and other polluting industries may be besieged by increasing numbers of claims upon them for compensation. This is already happening for Volkswagen, through their deception about carbon emissions from their diesel cars; and is likely to happen to ExxonMobil, who actively deceived the public about the reality of climate change over many years. The US coal giant, Peabody Energy, has already filed for bankruptcy.

If a financial crisis does occur and affects pensions, for example, the consequent loss of value (and potential income to pensioners) will come at a time when unprecedented numbers of people in the population reach pensionable age. The financial future of many thousands of people could thus be bleak.

This winter, there has been another catastrophic flooding event in the north of England, damaging many people's homes and putting even more pressure on the insurance industry, as well as causing £1.5 billion worth of damage to bridges, roads and other infrastructure.
Also, a recent investigation has shown that, of 20 zones earmarked by the UK government for the building of new homes, five were hit with alerts and warnings during these recent floods and storms.

Fig.63
Flooding in Appleby, Cumbria

Fig.64
The destruction of Pooley Bridge, Cumbria by flooding
From: www.bbc.co.uk

There is growing evidence that population growth and, more significantly, economic growth are the most important drivers in the increase in CO_2 emissions. Since 1970, emissions of CO_2 from fossil fuel combustion and industrial processes contributed to about 78% of the total GHG emission increase[78].

Professor Justin Lewis, in an article to The Independent[79] about a growing right-leaning bias in the BBC, argues that there is now a growing body of evidence suggesting that a model of permanent economic growth is of dwindling benefit to wealthy countries such as the UK. He cites research which shows that GDP growth is no longer linked to improvements in health or happiness, is environmentally unsustainable and stretches commodity choice far beyond the time we have available to us as consumers. He concludes that there is a serious debate about whether wealthy consumer economies should still rely on growth to generate prosperity.

Thus, there is an urgent need for the development of a new economic system. Just as the industrial revolution became the trigger for a change from agrarian economies to a market economy, **there needs to be a development of a new economic system that is triggered by climate change.** I do feel that the balance of the economy can be, and should be, adjusted to allow for the crisis that is heading our way. We can no longer continue to run the economy as if nothing is happening, with businesses maximising their short-term profit, with no heed for the wider damage that their activities are doing. **It is not "Business as Usual".** Those who currently make vast profits from a market economy, and who promote it as the only way forward, need to take stock and change their attitudes and behaviour. Our planet can no longer sustain the robbing of its resources, and the contamination of its atmosphere, in the name of progress (see chapter 3).

> "When the facts change, I change my mind. What do you do?"
>
> *John Maynard Keynes, Economist*
> *(1883-1940)*

Fig.65

Richard Douthwaite, in his book entitled *"The Growth Illusion: How economic growth has enriched the few, impoverished the many and endangered the planet"*[6,80], states that economic growth does not have the benefits that many (mainly economists) claim for it. He demonstrates that, if the rate of growth is fast enough, there may be increased business profits and extra jobs but little improvements in the lives of ordinary people. Douthwaite goes on to say that the benefits for businesses come at the cost of lower wages and reduced job security. He also comments that achieving growth through the global system exposes each of us personally, and the countries to which we belong, to much higher levels of financial and environmental risk than did the more nation-state-based economies of earlier generations. A full quotation from Douthwaite is given in a separate table (5).

Both Douthwaite[6] and Fletcher[81] (in *"Free Trade Doesn't Work"*, 2010) are of the opinion that economists have got things badly wrong, most of their theories being based on inappropriate mathematical equations. This thesis is further developed by Paul Krugman in the New York Times[82] in an article entitled, "How did Economists get it so Wrong?"

George Monbiot also addresses the issue of the mathematics being wrong in his article to The Guardian[82] and on his website. The article suggests that the calculations have given a false sense of reduction in the use of the earth's resources because they have failed to include goods purchased from abroad in the equation. Indeed, if you look at the UK alone, where carbon dioxide emissions apparently fell by 194 million tonnes between 2002 and 2012 (using the wrong calculations), the real figure cancels this out and gives in fact an increase in emissions, related to the commissioning and importing of goods.

The New Economics Foundation (NEF)[83] has done much work on alternative, more sustainable, economic systems. They are the UK's leading think tank on promoting social, economic and environmental justice. Their aim is to transform the economy, so that it works for people <u>and</u> the planet. However, one learns that they have received a large government grant to develop their work. Let us hope that they remain objective and are not influenced in their thinking by right-wing pressures.

They state on their website that:

"A strong national economy needs a flourishing network of local economies. These are shown to give resilience in times of crisis, but are consistently undermined by the sprawl of supermarkets and other chains - the kind of businesses that are most likely to up and leave in times of trouble. We should be ensuring that money stays in local communities rather than leaking out to distant head offices, and encouraging a range of diverse high streets rather than clone towns."

On Bankers and Banking, they state:
"A dysfunctional financial sector led us to the brink of disaster in 2008, and yet bank reforms aren't going far enough to tackle the root causes of the economic crisis. Our four big banks remain too big to fail, and continue to engage in the risky and unproductive activities that caused the crash. We need to establish a more stable, sustainable and socially useful banking system.

Jeremy Corbyn, in an article to The Times[84], stated that Britain must empower citizen suppliers and direct private investment into green technology. He believes that our weakened public services will not be able to cope with the consequences of drastic weather events, such as the floods in Cumbria (2015) and Somerset (2013-4). He states that we need carbon budgeting to be the centrepiece of trade and commerce, taking the planet back to sustainable levels of CO_2 emissions. Environmental politics must include people working in today's economy and decisions by government must not take us backwards but must instead invest in the huge opportunities that the low-carbon sector offers.

Colin Tudge presented a paper entitled "Economic Renaissance: Holistic Economics for the 21st century"[85] to a think tank at the Schumacher college in 2007. The think tank explored what the key components are of an economic system which would successfully achieve poverty elimination, climate sustainability and human fulfilment. What kind of economy do we need to protect ecosystems and people's livelihoods at the same time?

Professor Richard Murphy and Colin Hines wrote a report for discussion at the Paris 2015 Summit[86], which provides solutions for how new green measures might be funded. The suggestion is that some of the funds already allocated for Quantitative Easing to keep the financial system afloat by the European Central Bank (€7 trillion of new money being printed), should be allocated in the form of Climate QE to save the planet – a figure of €10 million a month is suggested. This could be used in the form of climate change bonds from the

European Investment Bank. These funds could then be directed to climate change programmes in Europe and in developing countries.

Fig.66
European currency

Other economists have suggested a different form of Quantitative Easing[87].

Some countries have introduced a carbon tax and, in some cases, this has been successful in lowering carbon emissions[88]. Sweden has been particularly successful, first introducing a carbon tax in 1991. Their economy has grown by 50% since that time[89] and their emissions of greenhouse gases have declined and been decoupled from economic growth. The OECD report[89], which looks in detail at a number of pollution factors showed that Sweden has cleaner air than most other countries in the world (OECD Environmental Performance Reviews Sweden 2014).

The experience of Australia has been different[90]. They introduced a carbon tax in 2012, whilst led by Prime Minister Julia Gillard under a coalition with the Greens. This act was extremely unpopular and was repealed two years later by Prime Minister Tony Abbot. Full details of how they implemented the carbon tax can be found in Wikipedia. The chart in figure 67 shows the carbon emissions falling during the carbon tax period (July 2012-July14) and then increasing again during July-November 2014, after it was repealed[90]. The decision to repeal the tax has led to Australia slipping from 10th to 13th in the RECAI list (see Table 4).

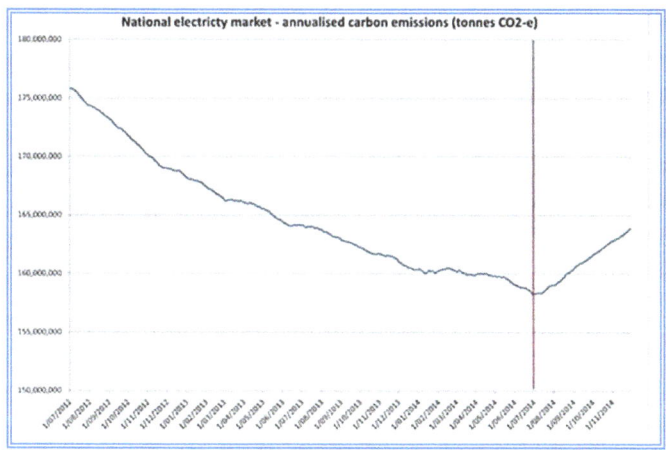

Fig.67
Carbon emissions in Australia before and after the carbon tax was repealed

A report from Climate News Network[91] indicates that Brazil, whose president Dilma Rousseff is at risk of being impeached, will dismantle their environmental protection laws after the impeachment, in defiance of commitments made in the Paris Agreement. It would appear that the leaders in the impeachment process are businessmen who would benefit from these laws being repealed. Here is yet another female leader, who introduced environmental safeguards, being undermined by big business. Brazil has just suffered the worst environmental disaster in its history: the bursting of a dam of toxic mud last year in Minas Gerais state. All animal and plant life was destroyed by this disaster, which is said to be the worst since Chernobyl.

Fergus Green, a Policy Analyst and Research Advisor to Professor Stern, recently published a paper, which gave evidence that a nation can develop and implement green policies, without damaging the economy[92]. The paper was issued to encourage discussion on the issues.

Professor Nicholas Stern has written a paper for Nature[93] in which he says that current economic models tend to underestimate seriously both the potential impacts of dangerous climate change and the wider benefits of a transition to low-carbon growth. He thinks that there is an urgent need for a new generation of models that give a more accurate picture and suggests that researchers across a range of disciplines (economics, engineering, science) work together to help those developing policy.

Other ideas have been to develop a system whereby the worst polluters have to foot the bill for cleaning up the damage they have wreaked on the planet.

Another group, linked to Feasta, have come up with a suggestion of capping the use of fossil fuels globally by introducing a global taxation system[94].

CapGlobalCarbon (CGC) aims to ensure that the aggregate global emissions from fossil fuels steadily decrease to zero. This would be achieved by a progressively tightening cap on fossil fuel extraction. Revenues from the extraction permits would be used to benefit the lowest consumers of fossil fuels. Such compensation could substantially alleviate poverty and reduce global inequality. By steadily and predictably reducing the global dependence on fossil fuels the process would also hasten a smooth transition to a zero-carbon economy.

Yet, despite all these well-argued documents and postings, in the UK at least, our present Government in the UK is rushing headlong into yet another era of austerity measures, based on the old economics, promoting growth and rewarding big business at the expense of the environment and the poorest in our society. And part of their strategy in taking this forward is to reduce spending on, and support for, green initiatives, aimed at reducing carbon emissions.

I am not an economist so I don't feel qualified to come up with a new system; there are others much better qualified than me to do this. All I have done here is to demonstrate how everything in this world is interconnected: ecosystems, the stratosphere, the industrial revolution and its continuum, population explosion, trading systems, weaponry and war, the rise of big business and bankers and the failed economics which they promote. This interconnectedness means that, if any one of these goes wrong or out of balance, then this will bring down all the others in a domino effect. Denial of this effect has only made things worse, with so much more to do to reverse the destruction.

It is interesting that the word *economy* has a similar root *(eco)* to the word *ecosystem* or *ecology*. I am told that *eco* comes from the Greek *oikos*, meaning 'house' or 'household'. I have shown in this book how ecosystems and the economy are inter-connected but what is needed is a new form of economy – or a new discipline – that appraises the needs of both through this inter-connectedness. Perhaps we should call it *ecosystomics* – a new form of the economy that provides for the human race, without damaging the ecosystems of the world.

Although I am not an economist, I feel I must put together some pointers (or suggestions) for those who do have the ability to construct such a system, making the changes necessary to have a balanced *green* economy. Shall I call it *Economy 6?* My readers may wish to add to it. It is my first venture into the new discipline of *ecosytomics*.

Suggestions for Economy 6

Some measures which might move us towards a new, balanced, green economy:

- For the introduction of greater incentive schemes to encourage businesses to develop, use and market greener technologies and to penalise those who don't. Examples of this could include: using and developing renewable forms of energy; phasing out motor vehicles which use petrol or diesel and introducing those that run on easily-accessible clean energy;

- Investing in research institutions which have the ability to develop innovative solutions to today's climate-change problems;
- Introducing legislation to reduce the use of the motor car, such as restricting the number of cars owned by each household, unless they run on clean energy;
- Phasing out coal-fired power generation and ending fossil fuel subsidies;
- Introducing a carbon tax on those companies who continue to use fossil fuels;
- Rebalancing the economy, so that the rich are not rewarded for irresponsible behaviour that adds to the carbon load;
- Setting targets, for meaningful reductions in carbon emissions by an early date, as suggested by Desmond Tutu in his petition (chapter 1) and ensuring that the calculations for this are correct;
- Phasing out nuclear power and nuclear weapons worldwide and re-channelling the money saved into the incentive-schemes and investments mentioned above;
- Proper funding of those institutions regulating the tax system, so that tax evasion and avoidance is properly penalised;
- Shifting the tax system to penalise those activities which need to be discouraged, such as greenhouse gas emissions and the accumulation of wealth;
- Banning certain household appliances and gadgets, which are not necessary and only add to the carbon load;
- Establishing a new institution, which will monitor the use of fossil fuels by companies and promote, and provide support for, the use of greener forms of energy;

- Encourage less air travel, by raising awareness about the damage this is doing to the planet and encouraging airlines to invest instead in technologies that do not damage the planet;
- Work globally with other partners to reduce deforestation;
- Re-balancing international trading systems, so that goods and animals are not transported unnecessarily across continents and seas, adding to the carbon load;
- Encouraging countries worldwide to be self-sufficient in terms of goods and resources, so that goods are not imported which can be produced internally;
- Re-think and re-balance entirely transnational trading systems;
- Work globally to find a better means of international co-operation in working jointly to reduce and reverse that damage that is currently being done to the planet;
- Encourage partnerships between local government and local cooperatives and social enterprises;
- Encouraging the setting up of local groups (3G groups), where individuals can meet together to share what they are doing to reduce their carbon emissions and to encourage each other to keep going with it, even if the majority of others are still in denial (3G stands for three generations – the amount of time we have left).

Some of the ideas above are already being worked on, and others are not about changing the economic system but about reducing carbon emissions, but I hope these are a starting point for others to add to, if we are really serious about taking meaningful anti-climate-change measures before it is too late.

Green Economy is not a new expression. It has been promoted by other groups, including the European Environment Agency, who produced the diagram in Fig.68:

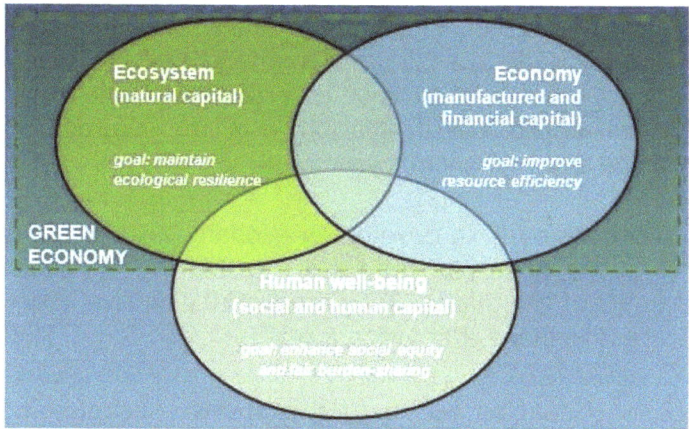

Fig.68

Their definition of a Green Economy is one that generates increasing prosperity while maintaining the natural systems that sustain us[95].

Their website goes on to say that:

1. Historically, the trend has not been towards green growth. On the contrary, economic expansion has imposed ever greater demands on natural systems – both in terms of the amount of resources that we extract or harvest, and the volume of emissions and waste that we expect the environment to absorb and neutralise.
2. As is increasingly understood, this cannot continue indefinitely: the environment has

natural limits in terms of how much it can provide and absorb.

There is also a body called "The Green Economy Coalition", which is a global network of organisations committed to accelerating a transition to a new green inclusive economy[96]. They believe the crisis we are in is profound and that piecemeal policy change is not good enough. They want to see deep-rooted transformation and the courage to forge a new economic vision.

Their vision is to develop an economy that provides a better quality of life for all, within the ecological limits of the planet. They are working on five action areas to make that vision a reality:

- Managing our natural systems – people and economies depend on nature for everything; until now our economies have not reflected that dependency;
- Investing in people – evidence shows that more equitable access to our natural resources benefits both people and planet; where communities have secure tenure and a say in decisions there are better outcomes for the environment as well as for the local economy;
- Greening high impact sectors – these sectors include food, housing and transport and their embedded energy needs – accounting for 63% of the global ecological footprint;
- Influencing financial flows – capital markets are dominated by large banks but smaller, values-based banks, which base their decisions on the needs of the people and the environment, have proved to outperform

traditional mainstream banks on all indicators, including financial ones;

- Measuring what matters – economic metrics, such as GDP and quarterly reports, tell us nothing about the resilience of an economy or business; some investors are asking for new metrics. Governments in Canada, Botswana and India are already working on this.

And just recently, the Indian Prime Minister, Narendra Modi, has invited 120-odd nations to join a new International Agency for Solar Policy and Application[97], aimed at helping poorer countries in the tropics develop solar power. India is investing $30 million to set up a headquarters and aims to raise a further $400 million. Modi has also written an article in The Economist (Gathering steam: The World in 2016, p70)[98], which outlines the basic precepts of India's economy, to include incentivising afforestation, setting ambitious targets for renewable energy, faster and more inclusive growth and eliminating poverty.

It is encouraging to know that many people across the globe are already looking to develop new economies that no longer threaten the sustainability of the planet. The agreement signed at the COP21 talks at the PARIS Summit also takes us in the right direction, even though fossil fuels are not mentioned in the wording of this agreement. The detail of this agreement will be discussed in chapter 8.

Fig.69
From: http://www.propostalavoro.com with permission

I have come to the end of describing the interconnected factors which have worked together to bring about a situation where the future of this planet is at risk. Many other people, apart from me, have realised this and are working in various ways to raise awareness and to lobby for change. However, I feel that far too many people, all over the world, do not understand the urgency of the situation and have not really curtailed their activities as a result. This is why I have written this book and tried to keep it simple.

The following chapters will look at how we may work towards global co-operation in a united effort to bring about a sustainable future. The final chapter looks at why, despite all the evidence to the contrary, there has been so much lethargy about doing something about it.

SUMMARY OF INTERCONNECTIONS WITH THE ECONOMY

CONNECTION	AS SHOWN BY
1. Industrial Revolution	1. Changes in economic systems occurred at the same time.
2. Market economies based on big business and multi-national companies	2. Further industrialization and the IR Continuum promoted and expanded.
3. Economic growth promoted as a good thing for national economies	3. Encourages greater industrialization which adds to the carbon footprint.
4. Market economy	4. Creates greater poverty.
5. Market economy	5. Adds to the amount of trading between countries and the global transportation of goods, all adding to the carbon footprint.
6. Market economy	6. Causes greater depletion of the earth's resources, encourages deforestation, leads to loss of habitat for many species and less absorption of carbon dioxide by forests.

7.	Market economy	7.	Encourages competition between nations, not the co-operation that will be needed to counter global warming.
8.	Catastrophic weather events	8.	Affect business stability, financial markets and the insurance industry.

Table 5

"A decade ago, few of us had any doubt about what growth was for. It was to lift people out of poverty and enable them to have a better quality of life. Political parties dressed these expectations up in different ways: the left would talk about growth leading to higher wages, improved social welfare, better hospitals, a lower pupil-teacher ratio and so on, while the right would stress greater profits and a wider range of choice.

But now much of the old confidence about the results of the growth process has evaporated...the only benefits many of us expect from economic growth are increased business profits and — if the rate of growth is fast enough — extra jobs.

....So why, since we know the benefits of growth have... hefty price tags attached, is it still considered so important to achieve it? One reason is that firms are constantly trying to lower their costs by introducing labour-saving technologies. Naturally these technologies cost jobs, so every year, unless the total amount of activity in the economy increases by about 3 per cent, unemployment will rise. As far as jobs are concerned therefore, national economies have to grow pretty quickly just to stand still.

The second reason our countries need growth is that between 15 and 20 per cent of their workforces are employed at any time on investment projects designed to expand their economies in the coming years. If growth fails one year, firms that invested but couldn't increase their sales in the flat market will find themselves with surplus capacity. This will cause them to cut any further investment plans they might have, throwing the people who would have built their new

173

factories, offices and shopping centres out of work. And since these newly unemployed people will obviously have less to spend, further jobs will be lost in other sectors of the economy. Consumer spending will fall even more, causing more job losses. In short, a downward spiral could develop leading to a serious depression. The possibility of this happening terrifies every government in the world to such an extent that they are prepared to do almost anything to ensure that growth carries on regardless of its social or environmental consequences......

In 1998, I conducted an Internet survey for almost 700 participants from over 50 countries. I had expected that it would take most of the seminar to reach some sort of agreement that, whatever growth might have achieved in the past, current growth was not benefiting ordinary people. Not at all. It took a bare 24 hours, so most of the seminary was spent discussing how the economic system could be altered to remove its need to grow.

CHAPTER 8
Global networks

Climate change, the loss of species, global warming, the increase in the human population, trading systems, the type of economy and poverty are all factors that affect every nation of the world in one way or another. If there is to be a change of direction, in order to save the planet and its inhabitants, it must happen on a global scale and include every country, or at least those countries which have industrialised. We need to get citizens across the world understanding the implications of climate change and industrialisation, so that they realise the need for urgent action and lobby their governments to make appropriate changes.

The most obvious organisation to initiate such a change of direction is (and has been) the United Nations.

The Efforts of the United Nations to reduce carbon emissions

Fig.70
The Rio Summit

In 1992, the United Nations Earth Summit, held in Rio de Janeiro, produced a document, called Agenda 21, which was a non-binding, voluntary-implemented action plan with regard to sustainable development. It provided an agenda for the UN, other multilateral organisations and individual governments around the world that could be executed at local, national or global level. The UN body proposed in Rio to take this forward was the UNFCCC (United Nations Framework Convention on Climate Change), whose director is currently Halldor Thorgeirsson. Since Rio, regular meetings have been held in different countries of the world, under the title of COP (conference of parties), the latest being COP21 in Paris. A further appraisal of the major COP agreements reached over the years is given in Table 6[99].

Fig.71

As the United Nations does have a role in addressing the issue of climate change, let's have a closer look first at how it functions and what it has achieved on climate change. The UN was first formed in 1945, as an intergovernmental organisation to promote international co-operation. The motivation for its formation came as a result of the Second World War, to prevent other similar conflicts from occurring. There were 51 member states initially and now there are 193, each country having one vote at deliberations of the General Assembly. The headquarters of the UN is in New York, with further offices in Geneva, Nairobi and Vienna. It is financed by contributions from its member states, the United Kingdom providing 5.19% of the total budget.

The UN currently operates through five principal bodies: the General Assembly (the main deliberative body); the Security Council (peace and security); the Economic and Social Council (ECOSOC) (for promoting international economic and social co-operation and development); the Secretariat (provides, information, studies and facilities needed); and the International Court of Justice. There are also various UN bodies, which have

177

particular functions: the World Bank; the World Health Organisation; the World Food Programme, UNESCO and UNICEF. The current General Secretary is the South Korean, Ban Ki-moon, whose term of office comes to an end during 2016.

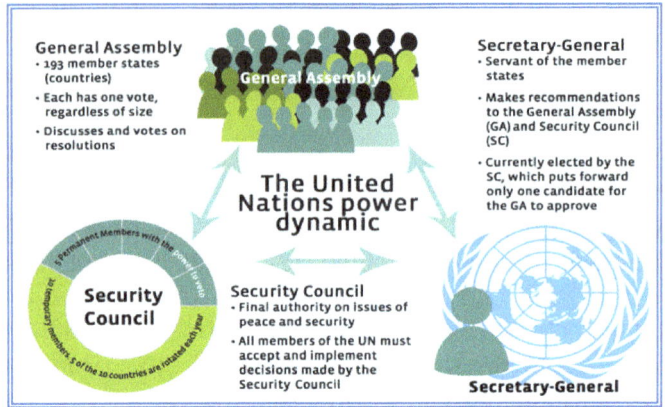

Fig.72
Structure of the United Nations

Many people are highly critical of the United Nations. Some commentators believe the organization to be an important force for peace and human development but others have called it ineffective, corrupt, biased and bureaucratic. I believe that, if the United Nations is to be taken seriously and respected, it needs to have more clout and to be reformed to be more inclusive. Its initiatives on climate change certainly need to be more decisive and more closely targeted. The problem is that, trying to get 193 or more nations to agree on one issue, is virtually impossible.

Agenda 21

The original Agenda 21 was divided into four sections:
- *Combating Poverty*;
- *II Environmental Issues*;
- *III Strengthening the role of major groups*;
- *IV. Means of Implementation.*

The "21" refers to the 21st Century and has been affirmed and modified at subsequent UN conferences. It is a 700-page document that was adopted by the 178 countries attending the 1992 conference. In 1997, the UN General Assembly held a special session to appraise the status of Agenda 21 and this has continued every 5 years since then. In 2012, at the United Nations Conference on Sustainable Development, the attending members reaffirmed their commitment to Agenda 21 in their outcome document called "The Future We Want". 180 leaders from nations participated.

The implementation of Agenda 21 was intended to involve action at international, national, regional and local levels. Some national and state governments have legislated or advised that local authorities take steps to implement the plan locally, as recommended in Chapter 28 of the document. The UN Department of Economic and Social Affairs' Division for Sustainable Development (ECOSOC) monitors and evaluates progress, nation by nation, towards the adoption of Agenda 21, as well as progress of the Millennium Development Goals (MDG), and makes these reports available to the public on its website. Europe has turned out to be the continent in which it was best accepted, as most European countries possess well documented

Agenda 21 statuses. France, for example has nationwide programmes supporting it, though there are opposition groups in this country, as there are in other countries.

In Africa, national support for Agenda 21 is strong and most countries are signatories. But support is often closely tied to environmental challenges specific to each country (such as desertification in Namibia) and there is little mention of Agenda 21 at the local level in the indigenous media. Agenda 21 participation in North African countries mirrors that of Middle Eastern countries, with most countries being signatories, but with little to no adoption at the local government level. Countries in sub-Saharan Africa and North Africa generally have poorly documented Agenda 21 status reports but South Africa's participation in Agenda 21 is similar to that of Europe.

Whilst the United States of America has been a signatory to Agenda 21, there is a strong business lobby, which opposes it on the grounds that it is bad for business. The Republican Party have stated that "We strongly reject the UN Agenda 21 as erosive of American sovereignty." Several state and local governments have considered or passed motions and legislation opposing Agenda 21, Alabama being the first state to prohibit government participation in it. Activists, some of whom have been associated with the Tea Party movement by The New York Times and The Huffington Post, have said that Agenda 21 is a conspiracy by the United Nations to deprive individuals of property rights. Interestingly though, in view of these opposition lobbies, the president of the USA, Barack Obama, recently participated in a TV documentary, in which he and David Attenborough discussed the issue of climate change and what needs to be done; he referred to a number of American initiatives to reduce carbon emissions.

Yet, despite the focus on environmental issues since the 1992 Rio Summit, global carbon emissions continue to rise, with petroleum, coal and natural gas being the worst culprits contributing to this increase.

Table 6

From Rio to Paris, UN milestones in the history of climate change discussions

International negotiations on climate change have been going on for over 20 years. In the meantime, the Earth has become hotter, wetter and wilder. Like scientists, the vast majority of governments now agree that urgent steps are needed to reduce our impact on global warming. So far, they have failed to sign up to a universal plan of action.

• **1992**: The United Nations Framework Convention on Climate Change (UNFCCC) was adopted during the Rio de Janeiro Earth Summit in 1992. It acknowledged the existence of human-induced climate change and gave industrialised countries the major part of responsibility for combating it – but without specifying how.

• **1997**: The adoption of the Kyoto Protocol in Japan in 1997 marked a milestone in international negotiations on tackling climate change. For the first time, binding greenhouse gas emissions reduction targets were set for industrialised countries, with obligations to reduce emissions by 5%. The protocol came into force in 2005, but was soon derailed by the failure of some of the world's biggest polluters, notably the US, to ratify it. As a result, other countries, such as Canada, Russia and Japan also pulled out. Another weakness of the Kyoto Protocol was that it exempted three countries, who were in the early stages of industrialisation (China, India, Australia) and now these are amongst the worst polluters. Protocol runs until 2020.

• **2007**: A longer-term vision was introduced by the Bali Action Plan in 2007, which set timelines for the negotiations towards reaching a successor agreement to the Kyoto Protocol, due to expire in 2012. It was expected that an agreement would be reached by December 2009.

• **2009**: Although the COP15 summit in Copenhagen, Denmark, did not result in the adoption of a new agreement, the summit recognised the common objective of keeping the increase in global temperature below 2°C. Furthermore, industrialised countries undertook to raise $100 billion per year by 2020 to assist developing countries in climate-change adaptation and mitigation, barring which poor countries had threatened to scupper any deal. That pledge became more tangible with the establishment of the Green Climate Fund in Cancún, Mexico, in **2010**.

• **2011**: Countries signed up to the Durban Platform for Enhanced Action (ADP), thereby agreeing to develop "a protocol, another legal instrument or an agreed outcome with legal force" applicable to all states that are party to the UNFCCC. This agreement was scheduled to be adopted in Paris and implemented from 2020.

At subsequent gatherings in Warsaw, Poland, in **2013** and Lima, Peru, in **2014**, all states were invited to submit their pledges towards reducing greenhouse gas emissions ahead of the COP21 summit in Paris.

The Paris Agreement

The objective of the latest, 2015 Paris COP21 conference was to achieve a legally binding and universal agreement on climate from all the nations of the world. I would like to see the UNFCCC go even further than this, to be involved in an internationally evidenced marketing initiative to demonstrate how urgent it is to take robust action to reduce carbon emissions. Just talking about it, and making gestures to the world press, is not going to achieve what is needed.

The COP21 talks in Paris set out more ambitious goals than many anticipated and was heralded, with much media attention, as an historic accord, though many hours had been spent in finalising the wording of this agreement, signed by 195 countries. Many believe that, in getting agreement, the main focus of the document was watered down. I have heard it said that certain oil-producing countries were the ones who caused the watering down of the Paris agreement, a similar action to that of Exxon Mobil, described in Chapter 3. The self-interest of powerful people in the world yet again holding back the actions required to really address the crisis that we all face.

The agreement included:
- Clauses to limit global warming to less than 2°C above pre-industrial levels and to endeavour to limit it to below 1.5°C;
- For countries to meet their own voluntary targets on limiting emissions between 2020 and 2030;
- For countries to submit new, tougher targets every five years;
- To aim for zero net emissions by 2050-2100;
- For rich nations to help poorer nations to adapt.

Fig.73
World leaders celebrating "an historic agreement" in Paris 2015

The agreement would come into force only after it had been ratified by 55 countries, who represented at least 55% of global emissions. If this target was exceeded, then the agreement would become operational in the same year.

A March 2016 report from the BBC indicated that the world's two biggest carbon emitters, the US and China (40% of emissions together), had produced a joint statement to say that both countries were ready to sign the agreement in April. Ban Ki-moon invited leaders to a signing ceremony in New York on 22nd April and expected 120 to turn up for this.

The move was not initially welcomed by some developing nations[100], led by an influential, Malaysia-based think tank who wanted to receive stronger assurances on finance, technology and compensation for damage from extreme weather before signing. Meena Raman of the Third World Network, was quoted as saying: *"It will be more advantageous to developing countries to wait this year and not rush into signing the Paris Agreement. Otherwise… we lose the political leverage that is critical to secure the necessary conditions that will enable developing countries to meet their obligations.*" Developing countries had therefore been advised not to attend or sign at the 22nd April ceremony.

However a list of 175 nations did sign on 22nd April 2016; those who signed are included on a UN website.
(http://www.un.org/sustainabledevelopment/blog/2016/04/parisagreementsignatures/)
as follows:

"List of Parties that signed the Paris Agreement on 22 April 2016[101]
The Paris Agreement will be open for signature by the Parties to the United Nations Framework Convention on 22 April and will remain open for signature for one year. This list contains the countries that signed the Agreement at the Signature Ceremony on 22 April:
Afghanistan, Albania, Algeria, Andorra, Angola, Antigua and Barbuda, Argentina, Australia, Austria, Azerbaijan, Bahamas, Bahrain, Bangladesh, Barbados, Belarus, Belgium, Belize, Bhutan, Bolivia (Plurinational State of), Bosnia and Herzegovina, Botswana, Brazil, Brunei Darussalam, Bulgaria, Burkina Faso, Burundi, Cabo Verde, Cambodia, Cameroon, Canada, Central African Republic, Chad, China, Colombia, Comoros, Congo, Costa Rica, Cote d'Ivoire, Croatia, Cuba, Cyprus, Czech Republic, Democratic People's Republic of Korea, Democratic Republic of Congo, Denmark, Djibouti, Dominica, Dominican Republic, Egypt, El Salvador, Equatorial Guinea, Eritrea, Estonia, Ethiopia, European Union, Fiji, Finland, France, Gabon, Georgia, Germany, Ghana, Greece,

Grenada, Guatemala, Guinea, Guinea Bissau, Guyana, Haiti, Honduras, Hungary, Iceland, India, Indonesia, Iran (Islamic Republic of), Ireland, Israel, Italy, Jamaica, Japan, Jordan, Kenya, Kiribati, Kuwait, Lao People's Democratic Republic, Latvia, Lebanon, Lesotho, Liberia, Libya, Liechtenstein, Lithuania, Luxembourg, Madagascar, Malawi, Malaysia, Maldives, Mali, Malta, Marshall Islands, Mauritius, Mauritania, Mexico, Micronesia (Federated States of), Monaco, Mongolia, Montenegro, Morocco, Mozambique, Myanmar, Namibia, Nauru, Nepal, Netherlands, New Zealand, Niger, Norway, Oman, Pakistan, Palau, Panama, Papua New Guinea, Paraguay, Peru, Philippines, Poland, Portugal, Qatar, Republic of Korea, Romania, Russian Federation, Rwanda, Saint Kitts and Nevis, Saint Lucia, Saint Vincent and the Grenadines, Samoa, San Marino, Sao Tome and Principe, Senegal, Serbia, Singapore, Slovakia, Slovenia, Solomon Islands, Somalia, South Africa, South Sudan, Spain, Sri Lanka, State of Palestine, Sudan, Suriname, Swaziland, Sweden, Switzerland, Tajikistan, Thailand, The Former Yugoslav Republic of Macedonia, Timor-Leste, Tonga, Trinidad and Tobago, Tunisia, Turkey, Tuvalu, Uganda, Ukraine, United Arab Emirates, United Kingdom, United Republic of Tanzania, United States of America, Uruguay, Vanuatu, Venezuela (Bolivarian Republic of), Viet Nam, Zimbabwe."

This is a very comprehensive list indeed and a significant achievement by Ban Ki-moon, though there are some notable absentees from the list (some oil-producing countries). Many people are now quite optimistic that there will be significant reductions in the use of fossil fuels and the subsequent carbon emissions. Others feel that the promised emissions' cuts are totally inadequate. In a review of the Paris agreement, Michael Le Page in the New Scientist (no. 3052) stated that he thinks time has nearly run out for limiting global warming even to 2˚C and he quoted from various scientists and leaders as follows:

"Emissions targets are still way off track, but this agreement has the tools to ramp up ambition, and brings a spirit of hope that we can rise to this challenge". Tony deBrum, foreign minister of the Marshall Islands.

"If we wait until 2020, it will be too late." Kevin Anderson, Climate Scientist at the Tyndall Centre in Manchester, UK.

"As for 1.5˚, it would take nothing less than "a true world revolution". We need renewable energy, nuclear power, fracking, zero-carbon transport, energy efficiency and housing changes. Even international aviation and shipping which were excluded from this report will need to be tackled". Piers Forster, University of Leeds.

I personally don't agree with this last person in terms of nuclear power and fracking, as I believe both to be dangerous cop-outs.

In a later, full length New Scientist Article, Michael Le Page[102] discussed the likelihood of countries being able to keep to the promises made. He reminded his readers that each signatory has to formally approve, or ratify, the deal in their parliaments and only five had so far done so: Fiji, Palau, Maldives, Marshall Islands and Switzerland an interesting group of countries most at risk of rising sea levels or melting ice. The Telegraph reported on 22nd April that there had been 15 ratifications[103].

Kimberley Nicholas, writing in the Scientific American (December 19th 2015)[104], discussed what is required to bring about the meeting of the 1.5° target. She quoted from an article in Nature Climate Change by scientists Rogelj and colleagues, that it will require "rapid and profound decarbonisation" from its **current 81% of fossil sources** in order to meet net zero carbon emissions as early as 2045 (recognised in the long-term goal in the Paris agreement to balance greenhouse gas emissions and removal). Further they had found that meeting the target would ultimately require actively removing carbon from the atmosphere, through means that have yet to be widely tested or implemented.

The December 2015 Newsletter of Tradable Energy Quotas (TEQs)[105] had the following statement about the Paris agreement:

"As the fallout continues, many of you may be confused by the outcome of the recent COP21 climate talks in Paris, variously reported as:
"A victory for all of the planet and future generations" ~ John Kerry, U.S. Secretary of State
"We did it! A turning point in human history!" ~ Avaaz
"10/10 for presentation, 4/10 for content" ~ Kevin Anderson, climate scientist
"A historic moment and positive step forward ... but not the legally-binding science and justice-

based agreement that was needed" ~ Friends of the Earth UK

"A sham" ~ Friends of the Earth International

"It's a fraud really, a fake" ~ James Hansen, climate scientist

"*Our leaders have shown themselves willing to set our world on fire*" ~ Naomi Klein, author/activist

"Epic fail on a planetary scale" ~ New Internationalist

"*The US is a cruel hypocrite. This is a deliberate plan to make the rich richer and the poor poorer*" ~ Lidy Nacpil, Asian People's Movement on Debt and Development

COP21: a clear win for political reality - a clear loss for every life form dependent on a liveable climate

The TEQs newsletter[105] continued:

"Our take is that when there is a fundamental rift between the physical reality of our changing climate and the political reality tasked with responding to this, this agreement - based on voluntary emissions pledges which even if met would mean more emissions in 2030 than today - is a clear win for political reality. In other words, a clear loss for every life form dependent on a liveable climate.

Sadly, it is not hard to identify the agendas of those hailing the Paris agreement as a great success. The whole conference has, in essence, been smoke and mirrors, distracting us from the real work of reintegrating human society with the reality that it depends on. As most impartial observers predicted, the UN have again failed to deliver an agreement that preserves the future of either humanity or the wider biosphere.

The Paris agreement is, in short, based on non-binding commitments to deliver on dodgy mathematics through the application of technologies that do not yet (and may never) exist."

Greenpeace have also criticised the Paris agreement[106], whilst applauding parts of it, such as setting 2018 as a review date. The main failure of the agreement, they feel, is that it failed the "justice test"; this relates to the human rights, where indigenous peoples affected by climate change are not given the protection they deserve. However, Greenpeace feel that what did not happen in Paris had already happened in Manila, where a human rights probe has been launched with the Human Rights Commission[106].

Thus, the challenge facing the world in 2016 is significant.   This has been reinforced by the excessive rain experienced in the north of the UK over the last few weeks, leading to extensive flooding, as well as in France and Germany and other extreme weather events in other parts of the world, such as the second strongest ever recorded tropical cyclone Winston which devastated Fiji.

I find it hard to reconcile these quoted comments with the "business as usual" attitude of our present government in the UK, a government which we will have to tolerate until 2020, unless something major happens in the next four years, to bring about an election.

In his book, *"Why are we Waiting"* (MIT Press), Professor Nicholas Stern[107], author of the Stern Review on the economics of climate change, sets out some of the goals that now face humanity in the 21st Century. The goals include:

1. The elimination of mass poverty and the risk of catastrophic climate change;
2. These goals are complementary;
3. The case for action is overwhelming because greenhouse gases stay in the atmosphere for centuries.

A recent research report in Science, and quoted in the Guardian[108], provides hope that carbon dioxide can be removed from the atmosphere by pumping it underground. The new research pumped CO_2 into the volcanic rock under Iceland and sped up a natural process where the basalts react with the gas to form carbonate minerals, which make up limestone. The researchers were amazed by how fast all the gas turned into a solid – just two years, compared to the hundreds or thousands of years that had been predicted. Juerg Matter, of the University of Southampton in the UK, led the research. Further research clearly needs to take place on this potential resolution to the problems we face.

But, are there other global networks we can call on to make a greater impact than that so far made by the UNFCCC?

Other initiatives

a) The Elders

In 2007, Nelson Mandela set up a group, called "The Elders"; it originally included elder states-people, such as Kofi Annan (now chairman of the group, former UN-Secretary-General and Nobel Peace Laureate), Archbishop Desmond Tutu (Nobel Peace Laureate and honorary elder), Aung Sun Suu Kyi (honorary elder until her election in 2012, Burmese pro-democracy leader), Ela Bhatt (India, pioneer of women's empowerment and grassroots development), Lakhdar Brahimi (Algeria, conflict mediator and UN diplomat) Martti Artisaari (Finland, Nobel Peace Laureate), Gro Harlem Brundtland (Norway, deputy chair, doctor who champions health as a human right), Fernando H Cardosa (Brazil – former president), Jimmy Carter (USA former president, Nobel Peace Laureate), Hina Jilani (Pakistan, pioneering lawyer and pro-democracy campaigner), Graça Machel (Mozambique, international advocate for women's and children's rights), Mary Robinson(first woman president of Ireland and former UN High Commissioner for human rights), Ernesto Zedilla (former president of Mexico who led profound democratic and social reforms).

Fig.74
A group of The Elders in 2010. From:
www.theelders.org

The Elders is an independent group of global leaders who work together for peace and human rights. The concept of the Elders originated from an idea from a conversation between the entrepreneur Richard Branson and the musician Peter Gabriel. The idea they discussed was simple: many communities look to their elders for guidance, or to help resolve disputes. In an increasingly interdependent world - a 'global village' - could a small, dedicated group of individuals use their collective experience and influence to help tackle some of the most pressing problems facing the world today? Branson and Gabriel took their idea to Nelson Mandela, who agreed to support it. With the help of Graça Machel and Desmond Tutu, Mandela set about bringing the Elders together and formally launched the group in Johannesburg in July 2007.

The Elders work strategically, focusing on work where they are uniquely placed to make a difference. One of their latest campaigns is for the UN, now over 70 years' old, to be adapted so that it is fit for purpose. They have four proposals on this (more details available on their website: http://theelders.org/un-fit-purpose):

- A new category of members;
- A pledge from permanent members;
- A voice for civil society;
- A more independent Secretary-General.

I believe that the issue of Climate Change is now so urgent that it may be too late to wait for a reform of the United Nations to tackle the issue more robustly. Perhaps a new body, independent of the United Nations, but respected globally, needs to take on the issue, cutting through all the bureaucracy that creates a climate of inaction on major issues.

Whether these proposals will bring about the changes necessary to generate greater respect and support for the United Nations, remains to be seen. However, the United Nations is the most obvious body to take forward the urgent imperative to work together with global co-operation to turn back the current surge of ever increasing carbon emissions and the devastating effects of climate change. Most of the concerns about climate change come from faith-based networks.

b) Christian-based organisations and networks have had much to say about the need for urgent action, as good stewardship of the earth is a major tenet of the Christian faith, as are the Jubilee principles of environmental restoration and fair allocation of wealth.

There is an ecumenical organisation, Operation Noah, with a seven-year plan to encourage Christians to work together to address climate change[109].

Recently, the Pope has issued an encyclical on climate change, which hasn't gone unnoticed[110].

For the Anglicans, Archbishop Desmond Tutu initiated a petition asking governments, and the United Nations, to set a renewable energy target of 100% by 2050[111]. Tutu, a Nobel peace laureate, who rose to fame for his anti-apartheid activism, said: "As responsible citizens of the world – sisters and brothers of one family, the human family, God's family – we have a duty to persuade our leaders to lead us in a new direction: to help us abandon our collective addiction to fossil fuels. We can no longer continue feeding our addiction to fossil fuels as if there were no tomorrow. For there will be no tomorrow."

There have also been statements published by:

- the Baptist Union[112]
- the Anglican Synod[113]
- the Methodist Church[114]
- the Quakers[115]
- and other ecumenical bodies, such as Christian Aid and Tear Fund.

A particular initiative is called "Eco-Church"[116], which encourages churches to switch their energy supplier to green forms of energy, with special rates being negotiated if a number of churches join the initiative[117] (called 'Big Church Switch'). Other bodies of Christians network to encourage individuals to reduce their personal carbon emissions, in various ways. A recent conference in Coventry, **"Hope in a Changing Climate"** provided much information to inspire hope, as many groups of Christians are working together, rather like the 3G groups I mentioned in Chapter 7, to reduce their personal emissions and to encourage their friends to do so as well. One speaker, a climate scientist, talked about efforts already underway to develop a plan for net zero (from the Paris agreement), with the aim of keeping 80% of fossil fuels **in the ground**. This included measurable actions to reduce carbon emissions per degree of warming.

There was also discussion about how churches might disinvest any funds they have with those companies who emit the most greenhouse gases, as well as taking action by joining the boards of such companies to influence their future direction. A similar action brought down the apartheid regime in South Africa. Indeed, it would appear that such an initiative is already underway through an organisation called Institutional Investors Group on Climate Change, based in London[118].

There has also been a Green Bible[119], which outlines text in green, which relate to environmental issues and teachings.

c) Other faiths
Other faiths making statements about climate change include Baha'i; Buddhism; Hindu; Islam; Sikh; Unitarian Universalist Association[120].

It is possible, therefore, that combined interfaith initiatives on climate change may have more impact on the activities of the global population than the United Nations has been able to do. Indeed, in 1995, at a conference in Japan on Religions, Land and Conservation, a declaration was made – *The Ohito Declaration*[121] – which stated ten spiritual principles:

1. Religious beliefs and traditions call us to care for the earth.

2. For people of faith maintaining and sustaining environmental life systems is a religious responsibility.

3. Nature should be treated with respect and compassion, thus forming a basis for our sense of responsibility for conserving plants, animals, land, water, air and energy.

4. Environmental understanding is enhanced when people learn from the example of prophets and of nature itself.

5. Markets and trade arrangements should reflect the spiritual needs of people and their communities to ensure health, justice and harmony. Justice and equity principles of faith traditions should be used for maintaining and sustaining environmental life systems.

6. People of faith should give more emphasis to a higher quality of life in preference to a higher standard of living, recognising that greed and avarice are root causes of environmental degradation and human debasement.

7. All faiths should fully recognise and promote the role of women in environmental sustainability.

8. People of faith should be involved in the conservation and development process. Development of the environment must take better account of its effects on the community and its religious beliefs.

9. Faith communities should endorse multilateral consultation in a form that recognizes the value of local/indigenous wisdom and current scientific information.

10. In the context of faith perspective, emphasis should be given not only to the globalisation of human endeavours, but also to participatory community action.

That declaration was made 11 years ago and, although people of faith make up the majority of the world's population, it is surprising that very little has been done so far to really get to grips with the damage to the environment and the planet that humans are responsible for. Perhaps the time has come for a new purposeful faith initiative. Table 7 gives a summary of the recommended actions proposed at the Ohiti Conference. Maybe it is time for all the religions of the world to take another look at it.

d) Other agencies
In chapter 7, I gave details of the **European Environment Agency** and the **Green Economy**

Coalition, both of which bodies are providing suggested frameworks for moving away from a market economy, which has been so damaging, to a **green economy**. Maybe either or both of these agencies can be reinforced to be the body to create more urgent change than the UNFCCC has done.

There is also **Forum for the Future**[122], an independent non-profit organisation, which works with business, government and other organisations to solve complex sustainability issues; they particularly focus on food and energy.

Table 7

Recommended Courses of Action made at the 1995 MOA International Conference on Religions, Land and Conservation, held in Ohito, Japan

1. We call upon religious leaders to emphasise environmental issues within religious teaching: faith should be taught and practised as if nature mattered.

2. We call upon religious communities to commit themselves to sustainable practices and encourage community use of their land.

3. We call upon religious leaders to recognise the need for ongoing environmental education and training for themselves and all those engaged in religious instruction.

4. We call upon people of faith to promote environmental education within their community especially among their youth and children.

5. We call upon people of faith to implement individual, community and institutional action plans at local, national, and global levels that flow from their spiritual practices and where possible to work with other faith communities.

6. We call upon religious leaders and faith communities to pursue peacemaking as an essential component of conservation action.

7. We call upon religious leaders and communities to be actively involved in caring for the environment to sponsor sustainable food production and consumption.

8. We call upon people of faith to take up the challenge of instituting fair trading practices devoid of financial, economic and political exploitation.

9. We call upon the world's religious leaders and world institutions to establish and maintain a networking system that will encourage sustainable agriculture and environmental life systems.

10. We call upon faith communities to act immediately, to undertake self-review and auditing processes on conservation issues on a regular basis.

A new body?

But some have no confidence in the United Nations and have no faith, so should we consider looking to form, or adopt, some of these other networks into a consortium, to bring about greater consensus about achieving measures to stop or reverse current trends? If so, how will these bodies be funded? Perhaps a global tax on all offending organisations would be apt, though probably unenforceable.

I leave this as a question for others in more influential positions than myself to answer, and/or implement, as necessary. Quite clearly there is a need for the nations of the world to stop seeing each other as competitors, rivals or enemies, for the desired results will not occur without **global co-operation**.

The Business World

There are businesses who are aware of the problems and who invest their profits in carbon reduction initiatives. These are showing the way for those large corporations who have been investing their profits in hiding the reality of climate change and in deceiving the public about their products and in paying so-called scientists to question the reality of climate change.

But so much more could be done, as it is often big business who has the financial resources to make a difference. Richard Branson played an active part in bringing The Elders together. As part of the business world (including the airline industry), which has brought us to the current dilemma, could he take a lead in getting business leaders together to understand, and rectify, what they have been responsible for, rather than burying their heads in the sand and continuing in their money-making at the expense of the planet. Recently, a podcast has been produced by Kyung-Ah Park[123] of Goldman Sachs on "The Business Case for Climate Action", as a result of attending the Paris Summit on behalf of this company. It is warming that some businesses are beginning to come up with strategies for the future.

How to make a global impact on the issues facing us

The writing of this book has changed my own attitudes and thinking. As a result I am no longer influenced by the rhetoric propagated by UK government and its economists to focus mainly on economic growth. For I know that, in promoting economic growth and redirecting funds to the business world, they are actually multiplying the effects of industrialisation and its by-products, which will further damage and destroy the ecosystems and atmosphere of this world.

I do not support initiatives to get involved in bombing countries far from our shores, in the name of national security, for I know that this all adds to the carbon footprint, as well as driving many indigenous people to flee their homes, adding to the thousands of refugees seeking new homes elsewhere.

But how can we reverse the centuries-old trend of global trade – of believing that free trade is a good thing? Trading systems and merchant cultures are at the root of all of the cycles I have described and I think I realised this when writing the *End Piece* to my first book.

There is still so much ignorance about the cycle of activities, described in the pages of this book. The general public tend not to see the urgency of the situation, or dismiss it as not their concern. If you have been influenced by the descriptions in the pages of this book, then use it to lobby for the changes that need to occur urgently.

CHAPTER 9
Bringing it all together and a way forward

So there we have it! A plethora of human activities which have put the harmonious cycles of our beautiful planet out of balance, leading to loss of habitat for many species, increasing global temperatures, climate change, extreme weather events, melting of the ice caps, raised sea levels, deforestation, acidification of the sea, space junk, accumulations of waste plastic and the threat of a mass extinction – all related to increasing carbon emissions, a process which may never be reversed unless urgent action is taken.

And, alongside of this, there has been the rapidly increasing human population, now seven times greater than at pre-industrial levels, leading to a multiplication of the destructive effects of human activity and loss of habitat for many species. Each of these activities has an inter-connectedness, which has led to a situation where a domino effect may take place, one factor triggering another factor, the total effect of which may make our planet unstable and uninhabitable in just three generations:

- Industrial revolution, which did not end pre-1900 but which continued with an ever-increasing momentum, through the IR Continuum, to the present time;
- Increasing human population, multiplying the effects of the IR;
- Changes to economies from local agrarian economies to market economies, which encourage further industrialisation and rewards

businesses who increase manufactured production;

- International and multinational trading patterns, adding to the IR Continuum and leading to local situations where more is imported than is exported, and politicians desiring to take action for more and more economic growth; such actions are counter-productive, adding to the carbon load;
- Greater divisions between the rich and poor in the world, leading to migration, unrest and wars, and with the rich contributing considerably more to climate change than the poor and with wars adding to the carbon footprint.

Fig. 75 attempts to show how all of these factors are interrelated and how each is contributing to ecological instability, both in its own right and by interaction with the others. For example, the increasing human population has a multiplying effect on all the others; weakening economies result in increased efforts to promote economic growth, which multiply the effects of industrialisation, trading systems and global travel; increasing affluence of the super-rich provides a multiplying effect through increased multi-national trading; poverty in some areas being related to deforestation in order to grow crops to survive, this has the effect of reducing the number of trees available to absorb carbon dioxide as part of the photosynthetic cycle; market economies exaggerate the effects of the industrial revolution and its continuum, as well as affecting trading systems;

greater unrest in the world, leading to wars, which add to the carbon load.

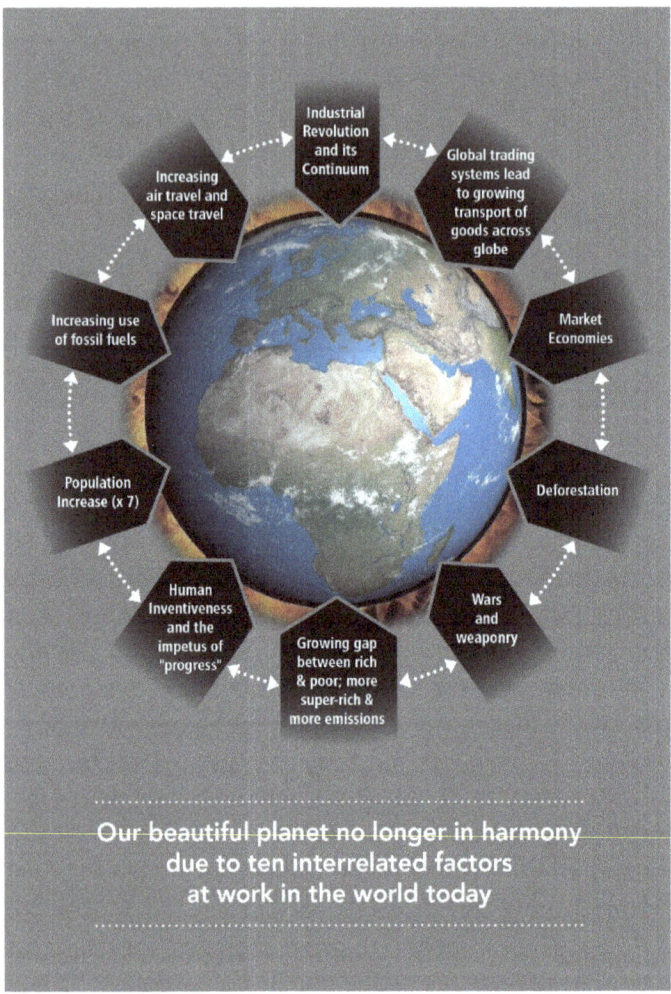

Industrial Revolution and its Continuum

Increasing air travel and space travel

Global trading systems lead to growing transport of goods across globe

Increasing use of fossil fuels

Market Economies

Population Increase (x 7)

Deforestation

Human Inventiveness and the impetus of "progress"

Growing gap between rich & poor; more super-rich & more emissions

Wars and weaponry

Our beautiful planet no longer in harmony
due to ten interrelated factors
at work in the world today

Fig. 75

I hope I have made a convincing case about the urgent need for change in the ways in which the global human population organises its affairs. To bring this change about needs **a complete re-think** by everybody, a complete change in the way in which we go about our normal lives and our business (see also Naomi Klein[8]).

As this book has unfolded, and during the writing of it, I have learnt so much myself – but this new learning has also opened me up to seeing things in a totally different light. It has been a revolution in my own thinking and responses. So, what started as a gut instinct has been transformed into an urgent imperative. I hope it does the same to you as well.

Many of the things that clutter up our lives, or make our lives more comfortable or exciting, have been produced at the cost of the planet.

So, some of the questions that have come to me, I will pose also to my readers:

- Can we continue to slavishly follow consumer trends? Buying the latest gadgets, regardless of whether they have been transported across the world, thus increasing carbon emissions?
- Can we continue to use our motor cars just to travel down the road to the shops or the school?
- Should we continue to import foodstuffs that can be produced in our own country by our own farmers?
- Can we continue to rob other species which share this planet with us, of their habitats?
- Can we continue to clutter up the space around our planet with redundant and unused space junk?

- Can we continue to fill our oceans with discarded, non-biodegradable plastic, which can also kill many marine species?
- Can we continue to support the free-trade movement, which feeds into further industrialisation and the IR continuum?
- Can we continue to give tacit support to a market economy, which rewards those companies and individuals who selfishly add to the carbon footprint of our planet?
- Can we continue to support those industries which make unheeding use of fossil fuels in order to make a profit for themselves?
- Can we continue to let the super-rich control most of the systems of the planet to feed their own greed, at the expense of the planet and of the poorest of the poor?
- Can we continue to use nuclear power and manufacture nuclear weapons, when there is no safe way to dispose of nuclear waste?
- Can we continue to go to war at the drop of a hat, when the carbon emissions released in such a war, add substantially to the carbon footprint of the planet?
- Can we continue to trade in weapons and spend vast amounts of money in producing them, when many millions of people across the world go hungry?
- Can we continue to allow the rich and multi-national companies to evade taxes, at the expense of supporting poorer nations to drag themselves out of poverty?
- Can we continue to ignore the comfortable relationship that our governments have with

the business world, which leads them to take decisions which support the business world, regardless of the damage they are doing to our planet and at the expense of the majority of the population?

- Can we continue to take long-haul holidays, travelling across the world, using airlines, which are one of the worst polluters of the atmosphere?
- Can we continue to give birth to babies, when the planet is already over-populated, with humans robbing many beautiful species of their habitats?
- Can we control or curb the results of human curiosity and inventiveness? Or should it be channelled into less world-destroying end-products, such as innovations to produce clean energy?
- Can we put sufficient pressure on our politicians to change direction regarding current economic thinking, the mathematics of which are flawed?
- Can changes in the economy be introduced in time to save the planet?
- What are the risks of changing trading practices on the economy?
- What if one country makes changes and is exploited by other, less scrupulous, nations?
- Will big business play ball?
- What about the rich super elite? Will they recognise the urgency of the situation and change their behaviour to a more altruistic approach?
- Is the idea of global co-operation to save the planet realistic?

So many questions have come to me and I am sure that other questions have come to you too as you have read this book. There are so many decisions we need to take as well, both as individuals and as nations and global citizens.

Some groups are looking at the issue of climate justice, in which reparation is made by the greatest polluters, to poorer countries whose way of life is severely affected by climate change. Indeed, this formed part of the COP21 Paris agreement.

Fig.76
© Joel Pett. Printed by permission.

Reasons for the lethargy

Of course, many people already realise and understand about the damage we have done to the planet, as a species, and many people are already taking action across the globe, but there seems to be a lethargy to make the significant changes needed, so I will address this too.

Part of the lethargy is, I think, due to the success of the big business-climate-change-deniers, who have influenced people to think that the scientists are wrong and that there is nothing to worry about. Naomi Klein[8] addresses this issue strongly in her book, "*This Changes Everything*", as she places most of the blame for the desperate situation we face today, firmly in the courts of the big corporate businesses. In her opinion, they have used their money, and lots of it, to establish a climate-denial movement, in which the credibility of the climate scientists is attacked and the seriousness of global warming is minimised. She identifies a strong right wing caucus, which sees those campaigning for climate action, as a group of left wingers who wish to establish their own political agendas on the rest of the world. They also see it as a new battle they must fight in much the same way as they fought communism during the Cold War. And they believe that they can use their $ millions to protect themselves from climate change disasters. In her view, they have little empathy for the victims of global warming and climate change, especially the poor in developing nations and island states; their attitude to such people is cruel and nothing short of racism.

Another reason for the lethargy is, I believe, that the whole concept of another mass extinction caused by climate change is too horrendous to think about and, in a way, unthinkable. Thus, people blank it out and just concentrate on their own lives and their normal agendas for the next few years. It is easier to do this than to institute, and campaign for, the major lifestyle changes that are needed to avert this crisis. And it is easier to label people who, like me, write books to raise the issues, as doom-mongers, greenies or left-wing loonies.

I have come across people who look at the greenness of the English countryside, stretching for mile upon mile and, looking at the lovely green foliage, they cannot take on board that this is likely to disappear and so, like others, they dismiss global warming as unlikely. The problem with this approach is that, we probably will see the countryside looking greener for a while, as plants and trees, in response to the increased carbon dioxide in the air, will produce more chlorophyll. This may have a minimal effect on atmospheric carbon dioxide. But the mind-set fails to acknowledge that ocean acidification is already taking place and rises in sea levels have already swallowed up some islands (five of the Solomon Islands, for example), that coral is bleaching and that the ice caps are melting at increasing rates. And that, whilst some areas are greening, other parts of the world are being ravaged and scorched by unprecedented temperatures and others suffering more and more wildfires and bush fires due to tinder-dry conditions (see also the quote from India at the end of chapter 10).

The scientist James Lovelock, who described the Gaia Hypothesis and who came up with methods to measure CFCs in the atmosphere, followed up his thesis with a warning[124.] He comments in his book that: *"... it seemed there was little understanding of the great dangers that we face. The recipients of climate forecasts, the news media, government departments, the financial market – normally as skittish as blushing teenagers – and the insurance companies all seem relatively unperturbed about*

212

climate change and continued with business as usual until their world, the global economy, almost collapsed."

Human Responses to warnings

One of the things that I find quite intriguing is how some people fail to take heed of warnings, a fire alarm for example. Whilst I get up, grab my things and run out of the door to the nearest fire exit, most people just carry on as if nothing had happened, as if the fire alarm were not ringing. The same thing happens on motorways, when warning messages urge you to slow down because there is some hazard ahead. Why is this? Of course, these could be false alarms but why take the risk? I can remember reading an article once about a tragedy when a ferry sank, drowning many people on board. It would appear that those who survived were the ones, who reacted immediately and made extreme efforts to get to the upper decks and the lifeboats. Are people unable to visualise a hazardous and different future? Why do we continue to live for the present even if it makes the future more risky - even fatal?

In an article in the New Scientist[125], Robert Gifford, a Canadian environmental psychologist looked at the psychological reasons why people have failed to take action on climate change. He came up with 33 reasons, which he grouped under certain headings. I'll attempt to give a short summary of them:

1. LIMITED UNDERSTANDING

Gifford believes that humans are far less rational than was once believed and gives a list of 10 reasons why humans are not acting on climate change. The reasons range through sheer ignorance, limited brain power, not knowing what to do about it, a lack of priority to climate change because it does not seem to be causing any immediate problems, hearing the message so often that we switch off to it (message numbness), not understanding the urgency of the situation, due to poor reporting, undervaluing distant and future risks, a tendency to over-optimism, a perception that climate change is a complex global problem, so people think their own behaviour will have little or no impact. Some have a fatalistic bias because they think nothing can be done, even by collective human action. People with doubts about the reality of climate change tend to read newspapers or listen to broadcasters which reinforce their convictions. Also, studies show that, when people view the time they have available to do something in monetary terms, they tend to skip acting in environmentally friendly ways. Some think they are unable to take climate-friendly action because they don't have the knowledge or skill and some claim they are unable to take certain actions, such as riding a bicycle or changing their diet.

Fig. 77
From:
http://www.durangobill.com/GwdLiars/GwdLiarsIndex.html

2. <u>IDEOLOGIES</u>

Gifford believes that there are four broad belief systems that inhibit climate-positive behaviour. These include a strong belief in capitalism, a tendency to justify the status quo, a belief that a religious or secular deity will not forsake them or that "Mother Nature will take a course that we mere mortals cannot influence" and a belief that technology will be able to solve all the problems.

This category of Gifford's has resonance with Naomi Klein's views, though he does not place it first, as she has.

3. <u>SOCIAL COMPARISON</u>

Gifford believes that, as humans are social animals, we will gravitate towards the choices of people we admire, so that, if they are climate change deniers, we will also deny that it is happening. He also believes that, if we see others not changing their behaviour, we will think, "Why should I change if they don't?" So this also leads to inaction about climate change.

Fig.78
From: Justin Bilicki, with permission

We buy things and spend money to make our lives more comfortable and some of these will not be climate-positive. They include financial investments, in a car, for example, or working in a fossil-fuel burning industry. Habit can also lead to repeating actions which increase climate change, in order to keep life more ordered and regular; people also have conflicting goals, values and aspirations, which do not always accord with climate friendly actions. People have strong aspirations to "get ahead" and their actions may compete with climate change goals, such as buying a larger house or car, taking an exotic holiday for example. This is a form of the consumer culture, which I mentioned in an earlier chapter. Gifford also believes that people get attached to a place and may thus oppose nearby wind farms (Nimbyism).

Fig.79
From : Joe Heller with permission

4. DISCREDANCE (OR DISAPPROVAL)

When people think ill of others, they are unlikely to believe what they say or take direction from them. For example, many people mistrust scientists, government officials or politicians, so do not take on board what they are saying. Some programmes have been introduced by government to encourage climate-friendly behaviour (such as solar panels at reduced costs) but are not considered by some to be generous enough. Large numbers of people in most countries do not believe that climate change is happening and so deny it; they are called climate change **deniers** and would include ordinary people as well as those with vested interests in using fossil fuels.

In chapter 3, I discussed the attractiveness of the concept of freedom and many people may struggle against what they consider will restrict their freedom. This includes big business, which strongly adhere to the free trade movement.

5. PERCEIVED RISK

Some people may consider that changing their behaviour and/or possessions is risky (eg buying an electric car, cycling instead of driving) or cost them too much or they may be afraid of being judged or teased by their peers for their choices.

6. LIMITED BEHAVIOUR

Most of us engage in some climate-friendly actions but these are not enough and may be just tokenistic. Others may make positive changes but these are cancelled out by other actions they take, which are not so climate-friendly.

It is helpful knowing the reasons why more action against climate change has not taken place but, in acknowledging these, we must also find ways to reduce their effect. In reading through them, I can find examples within my own behaviour amongst the lists, as well as in people I have discussed the issue with.   For example, I have found people with a strong sense of fatalism about it ("What will be will be"), as well as those who react as if its old hat: "We've heard it all before. What's new?"   I feel that perceived risks also feature very strongly and the government could do much more, by providing more generous subsidies for conversion to solar panels, for example, and by encouraging the motor industry to develop greener cars, which do not have perceived operational problems.

At the start of this book, I mentioned that it took me 22 years to begin to write it, after first becoming aware of the clouds of pollution hanging over each of the cities that I visited on my world trip in 1994. So, I have been part of the lethargy in a way that seems to hit most people to one degree or another. When I look back over those 22 years, I can see that I have been altering my behaviour in small ways to be more climate friendly, though like others, not by enough. Also, when I returned from my world trip in 1994, there were other imperatives for me to attend to, most of which have been described in my second book (*The Desert will Rejoice*). During that trip, I was introduced to many models of good social projects for working with the urban poor and marginalised and I became involved in developing or founding some new inner city projects. And I also had two other books to write – the story of my journey and the inspiration behind these inner city projects. So, global warming and climate needs went to the back of my mind. But they didn't totally disappear. Maybe a similar thing happens to others – we all lead such busy lives. Being too busy to take action about global warming may be another thing to add to Gifford's lists. But I am glad that I eventually became jolted into researching and putting together the evidence for this book.

And, for those who are still in denial after reading this book, I have just one thing to say "**JUST LOOK AT THE EVIDENCE**" and let it work on you, just as the clouds of hazy pollution I saw in 1994 eventually worked on me.

2015 has been the hottest year on record, this last winter too has been the wettest, with excessive rainfall leading to devastating floods in the north of England and elsewhere, causing £250 million worth of damage. The immediate reaction of people who have had their homes flooded is to accuse the government of not spending enough money on flood defences. This is important but, far more important is that they lobby government to do more to reduce carbon emissions nationally and to take a global lead to institute some of the changes necessary to avert global climate disaster. Just focussing on flood defences is an example of limited understanding (cognition) from Gifford's lists.

Is the idea of global co-operation to save the planet realistic?

This is a question I posed earlier in this chapter and it is worth looking at the difficulties in more detail. Global co-operation is the idea I have promoted throughout this book because I believe it is the only way to produce the kind of rapid changes in human activity that are needed if we are to save the world from destruction. We are all in this together, so the divisiveness promoted by some groups and countries is just not appropriate. The world is facing a crisis and we need to join hands and work together to solve it.

So, what are the factors which are likely to limit global co-operation? I list some of them below:

- The massive size of the global population;
- Differences in national priorities, ethos and cultures;
- Differences across the world in how climate change is affecting individual countries;
- Lack of trust between nations;
- Ideological differences;
- Other crises seem more important to address, such as terrorism, migration etc.;
- Risks to national economies;
- Fears that other nations will not do likewise;
- Fears of being left behind in trading competitiveness;
- Unwillingness to give up prestigious possessions, power and status.

Unless some of these factors are overcome, then global co-operation will not occur. They are all challenging but I do believe that the human intellect is capable of finding ways to take global co-operation forward. What is less likely to happen is to find the will to do it.

In the meantime

In the meantime a group of UK climate activists found themselves in the dock recently. The following is a post on Barbara Panvel's website "Antidote to doom and gloom" which describes what happened. The five activists had whitewashed the walls of the Department of Energy and Climate Change (DECC) and painted on them, in black: "The Department for Extreme Climate Change", to expose the department's hypocrisy.

The five activists, members of the Climate Change Action Group, were ordered to pay £340 each at Hammersmith Magistrates Court. The defendants, who represented themselves, did not dispute their presence at the scene or the actions attributed to them, but argued that they had a 'lawful excuse' under section 5 of the Criminal Damage Act.[126]

DECC was not fined.

Their letter, which was handed in to Energy Secretary Amber Rudd, made many powerful points. In a preamble, they declared:

"Climate change is not one in a number of issues to be addressed. A stable climate is a fundamental need on which the maintenance of our civilisation and the earth's abundant life relies. There will be no economy, health or security to speak of on the planet towards which we are currently heading".

Edited extract from list of actions June-Sept 2015, included in their letter, were the following:

In 2009 G20 countries, including the UK, pledged to phase out 'inefficient' fossil fuel subsidies. But on the 19 March 2015: George Osborne announces £1 billion worth of subsidies for North Sea Oil, on top of a whole series of previous measures, including support for further exploration:

16 June: The European Union says the UK is set to miss its EU target of generating 15 per cent of its energy (not just electricity) by renewable methods, despite being set one of the lowest targets of all EU countries.

17 June: On the evening of the Big Climate Lobby on the 17th June, when thousands met with their MPs to ask them to put climate as a priority, you announced the first of your 'cut-the-green-crap' policies, that new onshore wind farms (the cheapest form of renewable energy) will be excluded from a subsidy scheme from 1 April 2016, a year earlier than planned.

25 June: The UK says it will sell off up to 70% of its Green Bank, set up to lend money to risky green schemes such as wind farms that couldn't raise cash elsewhere. The sell-off means it may no longer focus on risky green schemes, and most of the profits will not go to taxpayers. By contrast, a similar US scheme is set to make $5 billion profit for taxpayers on $30 billion-worth of loans. Companies it helped include Tesla Motors, which paid back its loan early.

30 June: The Committee on Climate Change warns that the UK is not on course to meet targets after 2020. Its recommendations include taking action to encourage long-term investment in low-carbon energy, such as by extending existing short-term schemes to a 10-year timescale.

Ruth Jarman, one of the five members of the Christian Climate Action demonstration, who are deeply concerned about climate change and its impact on God's creation, the lives of people now the world over, and future generations, said:

"*We do not agree with today's judgement. The point of the law is to maintain justice, stability and order. Climate change threatens all these things so fundamentally that the law should be used to defend those who are trying to stop climate change, not those who are creating it. We think DECC should have been in the dock, not us. The department speaks fine words, but with its actions scuppers any possibility of global action to tackle climate change.*"

Michael Northcott, Professor of Ethics at the University of Edinburgh reminds us that without such acts in the history of the United Kingdom, the vote would not have been conferred on non-land owning citizens, nor on women and slavery, or forced child labour in our factories would not have ended. He said:

"*The actions of these protestors were a non-violent and peaceable way to expose the hypocrisy of current UK government energy policies. The UK has the potential still to lead the world towards the new sustainable energy economy that the climate crisis calls for and this type of action is essential to the democratic process in the UK."*

I believe that we will see many more actions like this, as the world in which we live gets more and more unstable.

CHAPTER 10
END PIECE TWO

I started this book by talking about my love of nature and how it had been present in me from a very early age and I shared that the fauna of this world have a very special place in my heart. And I saw how this special world of ours, originally so much in harmony and balance, was systematically being destroyed by the hand of man.

I cited an article by American scientists which argues that most of the life forms living on this earth will have become extinct in only three generations, with maybe humans becoming extinct at about the same time too. Hence the title of this book became: *"Three generations Left: Human Activity and the Destruction of the Planet"*. These scientists may be wrong about the dates and about the mass extinction but I believe that there is sufficient risk to rouse me into writing a book about it and attempting to show how other, apparently disconnected factors, have added to the risk.

The book has been targeted at the average person in the street, because I feel that the message in it has to become worldwide knowledge before serious actions are likely to be taken to reverse the destruction. There are too many vested interests to keep the status quo but the status quo will not be good enough any longer. We need radical system change.

The message in this book has not yet become universal knowledge. This was emphasised for me recently when I attended an anti-austerity workshop in Birmingham. A group of 50 or so thinking people had come together to discuss what might be the alternatives to the present Chancellor's austerity programme. We were split up into groups of five and given a poster to write down key factors that we considered to be important as alternatives to austerity. Then the groups were split up again, so that a different set of people was going through the same exercise. This happened four to five times. The thing that astonished me most was that each group seemed to have an entirely different list, though common themes did emerge, such as the environment. The other thing that astonished me was that the majority of people saw no link between anti-austerity and a green economy; indeed, many people did not know what a green economy was. Nobody mentioned loss of species and few were aware of the links between economies, trade, population increase, the industrial revolution, wars etc. that I have described in this book.

Whilst it was a shock to discover this lack of knowledge amongst thinking people, it has also been a spur for me to proceed to the publication of this book.

I have also been concerned that ordinary people, who are not particularly thinkers but who regularly read the red-top tabloids, have been strongly influenced by the lies that are, frequently and without conscience, spread across the pages and headlines of the daily papers that they read. I am sad that they have been so misled by a mixture of divisive rhetoric, scandal-mongering and fear-inducing falsification that is the situation we are living with today. How can people tell the difference between the truth and lies, when this is frequently being peddled to them by a frenzied media who gain from the tax breaks handed out to them through austerity economies, and who pander to the corporations because they want to receive advertising revenue from them to help them to balance their own books. They have no conscience about the lies that they propagate.

This is nothing short of corruption and it occurs, not only in today's media, but also in the business world, amongst the super-rich and in many politicians in power today, throughout the world. Several corrupt dictators have been brought down but others seem to get away with it because deceit and lies is their second nature and, if something is repeated often enough, people begin to believe in it as the truth. A good example of this was during the last two general elections in this country, when Conservative politicians repeated over and over that the Labour party were responsible for the 2008 recession and were weak on the economy. Many people believed this and voted the Conservatives into power as a result; the truth of the matter is that the 2008 recession was a world recession and the UK was not the only country to be affected by it. The recession was caused by banks being able to create too much money too quickly and used it to push up house prices and speculate on financial markets, so that debts became unpayable.

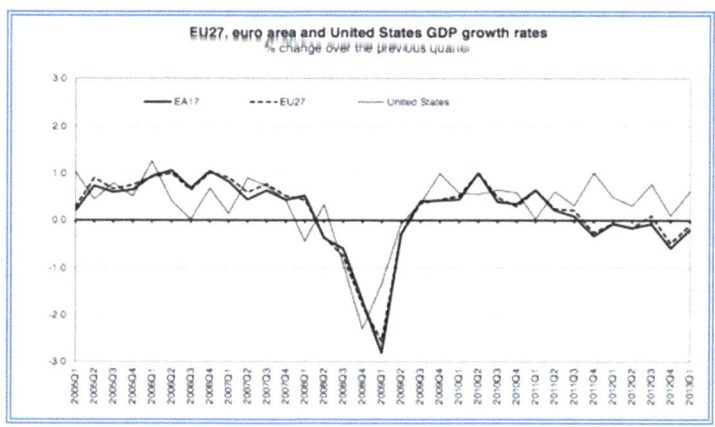

Figure 80 showing that the 2008 recession did not only occur in the UK but also in the Eurozone and the USA
(From:
https://www.economicshelp.org/blog/7157/economics/to-what-extent-did-eu-recession-cause-uk-recession/
Source ONS IHYQ)

From the same source as the figure above, is a bar chart of the UK economic growth during that same period, which shows that the economy had recovered before the 2010 general election began. And the truth of it is that, those who make this claim to be "strong on the economy" are actually not strong at all because the *market* economy as it stands at the moment (and as described in previous chapters), is actually destroying our beautiful world. This is not only occurring in the UK but in other countries too who have market economies. So, relentlessly pursuing a market economy is not the answer to the world's problems. It is positively dangerous.

The other point I want to make in this *"End Piece Two"* is about how power corrupts. When people get into a position of power and take rather dodgy actions from that position, and get away with it, never being taken to task by anyone, they gain in confidence to do it more and more, each time taking bigger and bigger risks. Thus, some politicians will go so far as to change their country's laws and constitutions to improve their chances of staying in power. This has happened in some African countries (e.g. Zimbabwe and now Uganda) and is currently happening in the UK, as constituency boundaries are being changed to improve the Conservative's chances of hanging onto power, as well as giving monetary handouts to Tory-run councils and squeezing the others. And their ability to do this is, of course, being fuelled by the super-rich.

Sometimes, political leaders who are in power for a long time become so confident that it goes to their head; some even develop a manic look in their eyes. You could say that they went power mad.

We have just had another budget in the UK issued by the present Chancellor, George Osborne and yet again, it is peddling this worn-out ideology of austerity measures, this time hitting disabled people even harder.

And, at the moment the media are in a frenzy about a forthcoming referendum to be imposed on the British public about whether to stay in the European Union or whether to come out, most of the frenzy being xenophobic or racist in nature. I fear this is a distraction. It is not the main issue we should be concerned about. If the earth on which we live is in danger, it is neither here nor there as to whether the UK is in Europe or not. The media, and the present government, is focussing on the wrong issue. Let us work together to ignore this distraction and to set a new agenda. An agenda to save our planet.

So, I will quote again from the "End Piece" to my first book:

"To reverse current trends, and to prevent the destruction of the world, there is an <u>urgent</u> need for co-operation between nations, in which the commonality of the human condition is stressed, rather than its diversity. Then, mankind might find a way to tackle global warming, to alleviate extreme poverty and to frustrate exploitation by the merchants."

Fig.81
From: http://www.catchnews.com/environment-news/cop-out-on-climate-change-will-paris-summit-achieve-anything-1448814544.html
Reproduced with permission from catchnews

And I will end by adding a quote (with permission) from Devinder Sharma in India, which is receiving much of the increase in global temperatures:

"It has now become even more obvious than before that the world we are living in has changed profoundly in the last five years. Every passing year is turning out to be hotter than the previous. It is just the middle of April but vast tracts of India are reeling under scorching heat with temperatures zipping past the 40 degrees mark. In 13 States, April temperature is higher by 8 degrees from the average. This will only intensify, as the season warms up.

India is on the boil, literally.

This is just the beginning of the summer months. In the next three months, before the monsoons set in, the heat wave is going to deadly. The Indian Meteorological Department (IMD) has predicted that the summer months this year will be warmer than normal across the country in all meteorological sub-divisions of the country. This year, unlike in the past, heat wave conditions are likely to hit more of central and north-western parts of the country. In fact, this is becoming quite visible with the hills facing very high temperatures.

I don't know why the IMD uses the word 'warmer' to describe sweltering heat conditions but shooting mercury has already taken a death toll of 130. If this is 'warmer' by IMD definition, I shudder to think what it would mean if it were to use the word 'hotter' instead?

Last year, 1,500 deaths from heat wave were reported from Andhra Pradesh alone.

Now, let us look at the rising graph of mercury. According to NASA, 2015 was the warmest year ever since it began to keep record. But a year earlier, in 2014, the world also lived through the warmest year till then. In other words, mercury has been rising with each passing year. And now, meteorological predictions globally point to a still warmer 2016. Let me add, India is not going to be an exception. The IMD too points to a deadly heat wave in the months ahead. Its predictions shows that 'all temperatures' maximum, minimum and mean for most sub-divisions from northwest India, Kerala from south India and Vidharbha from central India are likely to be above 1 degree C.

If you thought January was unusually warm this year, let it be known that February was still warmer. Globally, February 2016 was the hottest month known based on the long-term averages drawn. NASA had used the word 'shocker' to describe the unprecedented warming it measured for the month of February and warned of a 'climate emergency'. The average global temperatures in February were higher by 1.35 degree C. In India too, February was unusually warm this year with average temperature hike fluctuating between 1.5 degree and 2 degree.

But March has now turned to be the hottest. As per the World Meteorological Organisation (WMO) March has 'smashed' all previous records. Data compiled by Japan Meteorological Agency (JMA) shows that the March *temperature was higher by 1.07 degree, based on an average since 1891. Data released by NASA also shows that March temperatures have beaten the past 100-years records."*

We are now in mid-April and I can already feel the average temperatures creeping up. While we can survive, my thoughts go out to the 700 million people reeling under two consecutive years of drought. With wells almost dry and walking on a parched land they will now have to confront an unkindly hot sun. Some reports say wells have dried to a level in Marathwada not seen in past 100 years. Another report tells us that 133 rivers have dried in Jharkhand. To make matters worse, a BBC report indicated that the government might pipe Himalayan water and carry it all the way to the parched lands. After all, this is the surest way to add to GDP!

The relatively well-off in the cities, towns and suburbs have the facility to switch on an air-conditioner or an air-cooler but imagine the plight of majority population who have no other option but to survive under shade, be it at home or under the tree.

Water bodies have dried up. Many studies point to a steep fall in water levels in major reservoirs to the levels that are lowest in a decade. Reports of several rivers drying up are also pouring in, Tungbhadra in Andhra Pradesh being one of them. But while the media remained embroiled in the controversy surrounding IPL matches following the Mumbai High Court directive to shift them outside

Maharashtra, the nation has failed to focus on what is clearly a 'climate emergency'.

What should certainly be more worrying is that each year is turning out to be hotter than the previous. Quoting JMA, a report in The Guardian says: 'every one of the past 11 months has been the hottest ever recorded for that month.' The way the temperature is climbing every month, it seems the records will go on tumbling as we step into the future. Is this because of the climate change or not is something for the scientists and policy makers to conclude but as far as I am concerned the climate is already changing.

Can we do something? Yes, we can. There are already a number of stories of hope - of how ordinary people have made efforts and demonstrated the will to make a difference. Just to illustrate. From Anna Hazare's water harvesting techniques in the famed village of Ralegon Siddhi in Maharashtra to the tiny but forgotten village of Sukho-Majri tucked away in the Shivalik hills in Haryana, such examples are aplenty. This is just one way to minimize the impact. Several other alternatives and solutions have also been prescribed.

It's therefore high time to take a fresh look at what development means. Policy planning must shift to address the emerging issues linked to human survival at times of worsening climate. I am not sure whether the two-years of back-to-back drought followed by an unprecedented heat wave have given any jolt to policy planners. We seem to be simply *waiting for a normal monsoon to provide a succour, and wash away the dark realities."*

India is on a boil, literally. ABPLive.in April 16, 2016
http://www.abplive.in/blog/india-is-on-the-boil-literally
--

Posted By Devinder Sharma to Ground Reality at 4/26/2016 05:30:00 PM

REFERENCES

INTRODUCTION

1. Christine Parkinson, "*I will lift up my eyes*" (2002) New Generation Publishing, UK. p.301-302.

2. Jeremy Seabrook, Mark Tully and Molly Scott Cato, "Counting the Costs-1: an overview" (2005). https://neweranetwork.info/counting-the-cost-reports/

3. James Lovelock, "*Gaia*" (2009) Oxford University Press.

4. Winin Pereira and Jeremy Seabrook, *Asking the Earth: the spread of unsustainable development* (1996); The Other India Press, Goa, India.

5. Richard Douthwaite, *The Growth Illusion: How economic growth has enriched the few, impoverished the many and endangered the planet.* (1999) Green Books, Totnes, Devon.

6. Paul Rogers, Chances for Peace in the Second Decade – what was wrong and what we must do (2012) ORG Special Briefing, Oxford.

7. George Monbiot. www.monbiot.com and numerous articles in the *Guardian*.

8. Naomi Klein, *This Changes Everything* (2015), Penguin.

Chapter 1

9. Secrets of our Living Planet (2012), presented by Chris Packham, BBC, London DVD.

10. James Lovelock, *Gaia – A new look at life on earth* (1979), Oxford University Press.

11. Gerardo Ceballos, Paul Ehrlich, Anthony D. Barnosky, Andres Garcia, Robert M. Pringle and Todd M. Palmer. Stanford Report (June 19th 2015) Science Advances, California, USA.

12. Anthony Costello, Richard Horton. Health and Climate Change (June 23rd 2015) *Lancet* Commission.

13. NIEHA Interagency Working Group (April 2010) A Human Health Perspective on Climate Change.

14. Matt McGrath (2015, 9th November) Warming set to breach 1°C threshold. BBC: www.bbc.co.uk/news/science-environment-34763036.

15. Damian Carrington, *Guardian*, 15th April 2016. March temperature smashed 100-year global record. www.theguardian.com/environment/2016/April/15/march-temperature-smashes-100-year-global-record

16. http://fortheloveof.org.uk/i-wish/?org=rspb&sourcecode=EMLGE10026&contact=152968016SUS&utm_source=Febenews&utm_medium=Email&utm_campaign=enews.

Chapter 2

17. http://www.statista.com/statistics/300305/number-of-new-car-registrations-in-the-united-kingdom/

18. Vehicle Data, Society of Motor Manufacturers and Traders (SMMT) (2016) www.smmt.co.uk. Registrations – Cars.

19. Jeff Cobb (2014 and 2016) Top 6 Plug-in vehicle Adopting Countries. In: www.HybridCars.com

20. Peter Dockrill (13th April 2016) In: ScienceAlert. http://www.sciencealert.com/the-netherlands-is-making-moves-to-ban-all-non-electric-vehicles-by-2025

21. W. Pereira and J. Seabrook (1996), *Asking the Earth; the Spread of Unsustainable Development*. The Other India Press.

22. www.followgreenliving.com/deforestation-uttarakhand-disaster-rtr/

23. Adam Smith (1776), *An Inquiry into the Nature and Causes of the Wealth of Nations*; W. Strahan and T.Cadell, London.

24. Data and charts on electricity generation from www.carbonbrief.org

25. Damian Carrington, *Guardian* (24th September 2015): Renewable energy outstrips coal for first time in UK electricity mix.

26. Jonathan Watts, *Guardian* (3rd December 2015): Uruguay makes dramatic shift to nearly 95% electricity from clean energy.

27. Dutch student's cheap solution for clearing plastic rubbish from oceans: https://www.facebook.com/sydneymorningherald/videos/10154036592931264/

28. Molly Scott Cato (2005) Counting the Costs Report-1. http://neweranetwork.info/counting-the-cost-reports/counting-the-costs-1-an-overview/

29. H. Damon Matthews, Quantifying historical carbon and climate debts among nations. April, (2015) Nature Climate Change: http://www.nature.com/nclimate/journal/vaop/ncurrent/fig_tab/nclimate2774_F1.html

Chapter 3
30. Paul Rogers (2012) ORG Special Briefing, Oxford Research Group. Chances for Peace in the Second Decade – What is going wrong and what we must do.
And:
https://www.youtube.com/watch?v=rDg4F769ciw

31. Financial Times (2016) London Breaks 2016 air quality rules, 175 hours into new year. http://www.ft.com/cms/s/0/f7dbaa86-b617-11e5-b147-e5e5bba42e51.html#axzz43iwVXNYa

Chapter 4
32. Shannon Hall, Exxon Knew about Climate Change almost 40 years ago. (October 25th 2015), Scientific American: http://www.scientificamerican.com/article/exxo

n know-about-climate-change-almost-40-years-ago/

33. Rupert Neate (23rd March 2016) *Guardian:* "Rockefeller Family Charity to withdraw all investments in fossil-fuel companies."

34. *Guardian* (20th November 2013) Just 90 Companies caused two-thirds of man-made global warming emissions. http://www.theguardian.com/environment/2013/nov/20/90-companies-man-made-global-warming-emissions-climate-change Study by Richard Heede.

35. Greenpeace, 20th November 2013: http://www.greenpeace.org/international/Global/international/briefings/climate/2013/Carbon-Major-factsheet.pdf

36. Greenpeace Philippines: http://globalnation.inquirer.net/128747/carbon-majors-primary-drivers-of-climate-change-identified.

37. Car makers accused of 'obstructive' lobbying over emissions (November 2015) Exaro News. http://www.exaronews.com/search/node/InflueNceMap

38. Carbon Brief, The Global Oil Trade (2015) http://www.carbonbrief.org/interactive-how-the-global-oil-trade-is-changing

39. Ian Fletcher (2010) *Free Trade Doesn't Work*, US Business and Industry Council, Washington, D.C., USA.

40. Caroline Lucas (2002) Stopping the Great Food Swap: Relocalising Europe's Food Supply.

41. Colin Hines, Localisation: a Global Manifesto, Earthscan (2000).

42. Rianne ten Veen (2011), Global Food Swap, Counting the Costs -4;
www.neweranetwork.info

43. Local Futures Action Paper, Climate Change or System Change? (2015)
http://www.localfutures.org/wp-content/uploads/Climate-Change-or-System-Change-1.pdf

44. Colin Tudge (2016) *Six Steps back to the Land: why we need small mixed farms and millions more farmers*. Green Books, Cambridge.

45. Richard Douthwaite, *The Growth Illusion: How economic growth has enriched the few, impoverished the many and endangered the planet* (1999) Green Books, Totnes, Devon.

46. Muhammud Yunus, *Creating a World without Poverty* (2007) PublicAffairs, Perseus Group, USA.

47. Declaration on Green Growth (2009 and 2011) OECD.
https://www.oecd.org/env/44077822.pdf and http://www.oecd.org/greengrowth/48012345.pdf

40. Paul Tudor Jones II, Why we need to rethink Capitalism.
https://www.youtube.com/watch?v=dvJSK4viVMs

49. Lee Williams, *Independent* (6th October 2015) What is TTIP? And six reasons why the answer should scare you.
http://www.independent.co.uk/voices/comment/what-is-ttip-and-six-reasons-why-the-answer-should-scare-you-9779688.html

50. Centre for Economic Policy Research (CEPR) and Robert Schuman Centre for Advanced Studies: The Global Trade Slowdown: A New Normal? Edited by Bernard Hoekman.
http://voxeu.org/sites/default/files/file/Global%20Trade%20Slowdown_nocover.pdf

51. https://www.worldbank.org/content/dam/Worldbank/GEP/GEP2015a/pdfs/GEP2015a_chapter4_report_trade.pdf

Chapter 5
52. Thomas Malthus: An Essay on the Principle of Population (1798), pamphlet published in London under the name of Joseph Johnson – see details in Wikipedia.

53. *Guardian* (29th October 2015): China ends one-child policy after 35 years.

54. Hans Rosling (2015), Don't Panic – the facts about population:
www.youtube.com/watch?v=jbkSRLYSojo

55. Humanity's Last Stand (2015). In: Open Minds, a magazine of the Open University.

56. Paul Ehrlich and Ann Ehrlich (2008). *The Dominant Animal: Human Evolution and the Environment*. Island Press/Shearwater Books, Washington, USA.

57. Paul Rogers (Sept.2012) Chances for Peace in the Second Decade – What is going wrong and what we must do. ORG Special briefing Series for the Oxford Research Group.

58. Andrew Sayer (2014) *Why we can't afford the rich*. Policy Press. And:
https://vimeo.com/70068180.

59. Oxfam figures of richest 1%:
http://www.oxfam.org.uk/blogs/2015/01/riches t-1-per-cent-will-own-more-than-all-the-rest-by-2016

60. Helena Norberg Hodge, *The Economics of Happiness*, 2011.
https://www.youtube.com/watch?v=4r06_F2FIKM

61. The industrialisation for farming:
http://www.wwf.org.uk/filelibrary/pdf/ag_in_th e_eu.pdf
Legislation about this:
https://www.gov.uk/guidance/countryside-hedgerows-regulation-and-management).

62. George Monbiot (27th May 2015) A Pre-History of Violence. *Guardian* and also on
www.monbiot.com

63. Carla Garnett (2008) NIH Record, Volume LX No. 15. A Review of Calhoun's experiments on over-crowding with rats (Scientific American,1962).

64. Richard Douthwaite (1999) *The Growth Illusion*, Green Books.

65. Karen Jeffrey & Juliet Michaelson, New Economics Foundation (2015) Five Headline Indicators of National Success.
http://www.neweconomics.org/publications/entry/five-headline-indicators-of-national-success

Chapter 6

66. The Independent (Sunday 20th Sept. 2015) Ministry of Defence condemn army general behind Jeremy Corbyn 'mutiny' threat.
http://www.independent.co.uk/news/uk/politics/ministry-of-defence-condemn-army-general-behind-jeremy-corbyn-mutiny-threat-10510353.html

67. Global Warming and the Iraq War (2008) Climate & Capitalism, March 19th.
http://climateandcapitalism.com/2008/03/19/global-warming-and-the-iraq-war/

68. Paul Rogers (Sept.2012) Chances for Peace in the Second Decade – What is going wrong and what we must do. ORG Special briefing Series for the Oxford Research Group.

69. John Greenberg (24th June 2014) Iraq war dollars could have ended world hunger for 30 years.
http://www.politifact.com/punditfact/statements/2014/jun/24/facebook-posts/facebook-meme-iraq-war-dollars-could-have-ended-wo/

Chapter 7

70. Pat Conaty (November 2015) A Collaborative Economy for the Common Good.
http://wales.coop/publications/

71. Jonathon Porritt (2015) The Coalition Government 2010-2015; The Greenest Government Ever: By no stretch of the Imagination.
http://www.theecologist.org/News/news_analysis/2912348/the_greenest_government_ever_by_no_stretch_of_the_imagination.html

72. Michael Le Page (2015) "Ungreen and not-so-pleasant land". *New Scientist* No. 3042.

73. Ben Warren (Ed.) (2015) Renewable energy country attractiveness index (RECAI) Issue 43.
http://www.ey.com/Publication/vwLUAssets/Renewable_Energy_Country_Attractiveness_Index_43/$FILE/RECAI%2043_March%202015.pdf

74. Donald Braben (12 Sept. 2015) *New Scientist* No. 3038, p24-25.

75. James Bloodworth (10th June 2014) *Independent*. "It's time to bust some myths about benefit fraud and tax evasion."
http://www.independent.co.uk/voices/comment/its-time-to-bust-some-myths-about-benefit-fraud-and-tax-evasion-9520562.html

76. Heather Stewart, *Guardian* (21st July 2012), Wealth doesn't trickle down, it just floods offshore, research reveals.
https://www.theguardian.com/business/2012/jul/21/offshore-wealth-global-economy-tax-havens

77. Mark Carney (2015) Breaking the tragedy of the horizon – climate change and financial stability. Speech given at Lloyds of London 29th September 2015.

78. QUNO Climate Change Science Report (2014) http://www.quno.org/timeline/Human-Impacts-of-Climate-Change

79. Justin Lewis (2014) How the BBC leans to the right. The Independent, 14th February 2014. http://www.independent.co.uk/news/uk/politics/bbc-accused-of-political-bias-on-the-right-not-the-left-9129639.html

80. Richard Douthwaite (1999) *The Growth Illusion*, Green Books, Totnes, Devon.

81. Ian Fletcher (2010) *Free Trade Doesn't Work*, US Business and Industry Council, Washington, D.C., USA.

81. Paul Krugman (2009) How did economists get it so wrong? New York Times, 2nd Sept. 2009.

82. George Monbiot (2015) *Guardian*, 24th November 2015. Consume More, conserve more: Sorry but we just can't do both.

83. New Economics Foundation: http://www.neweconomics.org/issues/entry/banking-finance1

84. Jeremy Corbyn (2015) *The Times*, 9th December 2015.

85. Colin Tudge (2007) *Economic Renaissance: Holistic Economics for the 21st Century.* Schumacher College think tank, published by Green Books.

86. Richard Murphy and Colin Hines (2015) Finance for the Future: Climate QE for Paree. www.financeforthefuture.com

87. Quantitative Easing: https://thomasattwood.wordpress.com/

88. Carbon tax: http://www.carbontax.org/where-carbon-is-taxed/

89. OECD Environmental Performance Review: Sweden 2014. https://issuu.com/oecd.publishing/docs/sweden_ar_brochure_web

90. Carbon tax in Australia: https://en.wikipedia.org/wiki/Carbon_pricing_in_Australia
http://www.carbontax.org/where-carbon-is-taxed/

91. Jan Rocha, *Climate News Network* (May 12th 2016), Brazil Prepares to Roll back Green Laws. http://climatenewsnetwork.net/18257-2/?utm_source=Climate+News+Network&utm_campaign=cf92aa3172-Brazil_impeachment5_12_2016&utm_medium=email&utm_term=0_1198ea8936-cf92aa3172-38687953&ct=t(Brazil_impeachment5_12_2016)&mc_cid=cf92aa3172&mc_eid=3f6f7682e0

92. Fergus Green (2015) Nationally self-interested climate change mitigation: a unified conceptual framework. Working

Paper 224, Centre for Climate change Economics and Policy and Working Paper 199, Grantham Research Institute on Climate Change and the Environment.

93. Nicholas Stern (2016) *Economics: Current Climate Models are grossly misleading*. Nature News 530, 407-409.

94. Global Taxation System:
www.capglobalcarbon.org

95. Definition of a Green Economy:
http://www.eea.europa.eu/themes/economy/intro

96. The Green Economy Coalition:
http://www.greeneconomycoalition.org/know-how/informal-economy-threat-or-driver-green-economy

97. International Agency for Solar Policy and Application. *New Scientist*, 3030.

98. Narendra Modi (2016) *The Economist*; "Gathering Steam: the World in 2016", p.70.

Chapter 8

99. From Rio to Paris: UN Milestones in the history of climate change discussions. Taken from Wikipedia and summarised for this book in Table 7.

100. Ed King, Climate Home (29th March 2016) Developing nations urged to boycott Paris Agreement signing.
http://www.climatechangenews.com/2016/03/2

9/developing-nations-urged-to-boycott-paris-agreement-signing/

101. Signatories to the Paris agreement:
http://www.un.org/sustainabledevelopment/blog/2016/04/parisagreementsingatures/

102. Michael Le Page (2015) *New Scientist*, No. 3052, p8-9. Will Paris deal save our future? And: *New Scientist*, No. 3060, (2016) Signed, sealed……..Undeliverable?

103. Reuters (22nd April 2016) *Daily Telegraph*: Record signatures in historic Paris climate deal.
http://www.telegraph.co.uk/news/2016/04/22/record-signatures-in-historic-paris-climate-deal/

104. Kimberley Nicholas (2015) *Scientific American*, 19th December 2015. Limiting Global Warming will be hard but hardly impossible.

105. Tradable Energy Quotas (TEQs) Newsletter, December 2015.
https://neweranetwork.info/category/david-fleming/

106. Greenpeace:
http://www.greenpeace.org/international/en/news/Blogs/makingwaves/paris-agreement-expectations-COP21-wrapup/blog/55112/?source=em&subsource=20151218floem01&utm_source=gpeace&utm_medium=em&utm_campaign=20151218floem01

107. Nicholas Stern (2015) *Why are we waiting: the Logic, Urgency and Promise of tackling climate change*, MIT Press.

100. Damian Carrington (2016) *Guardian* 9th June 2016: CO2 turned into stone in Iceland in climate change breakthrough, quoting from research report by Matter et al in *Science*, 352, issue 6291, pp 1312-1314: Rapid carbon mineralization for permanent disposal of anthropogenic carbon dioxide emissions.

109. Operation Noah:
http://www.arcworld.org/downloads/Interfaith-UK-OperationNoah-7YP.pdf.

110. Pope's encyclical on Climate Change (2015)
http://cafod.org.uk/content/download/25373/182331/file/papa-francesco_20150524_enciclica-laudato-si_en.pdf and
http://s3.documentcloud.org/documents/2068632/climate-change-and-the-common-good.pdf

111. Desmond Tutu's climate petition tops 300,000 signatures
http://www.theguardian.com/environment/2015/sep/10/desmond-tutus-climate-petition-tops-300000-signatures

112. Baptist Union statement on climate change
http://www.baptist.org.uk/Articles/422769/Climate_Change_and.aspx

113. The Anglican Synod statement on climate change
https://www.churchofengland.org/media-centre/news/2015/07/urgent-action-needed-on-climate-change-urges-synod.aspx

114. The Methodist Church statement on climate change
http://www.methodist.org.uk/mission/climate-change

115. The Quakers' statement on climate change
http://www.quakerearthcare.org/article/shared-quaker-statement-facing-challenge-climate-change/

116. Eco-Church: http://ecochurch.arocha.org.uk/

117. Big Church Switch:
https://www.bigchurchswitch.org.uk/

118. Institutional Investors group on Climate Change:
http://www.top1000funds.com/tag/institutional-investors-group-on-climate-change/

119. The Green Bible – Harper Bibles, published by Harper Collins Publishers:
www.harpercollins.com/9780061627996/the-green-bible

120. The statements of other faiths on climate change:
http://www.preachin.org/wp-content/uploads/2016/01/cop_statement_formatted_-_website_version.pdf
http://fore.yale.edu/files/Buddhist_Climate_Change_Statement_5-14-15.pdf
http://safcei.org/wp-content/uploads/2016/01/A-Hindu-Declaration-on-Climate-Change-.pdf
http://islamicclimatedeclaration.org/islamic-declaration-on-global-climate-change/

http://www.interfaithpowerandlight.org/resourc
es/religious-statements-on-climate-change/first-
sikh-statement-climate-change/
http://www.uua.org/statements/threat-global-
warmingclimate-change

121. Ohito Declaration on Religions, Land and
Conservation (1995):
http://www.xiao-
en.org/cultural/life.asp?cat=54&loc=zh-
cn&id=1021

122. Forum for the Future, Annual Report 2014,
System innovation in action: our progress
and achievements.
www.forumforthefuture.org

123. Kyung-Ah Park, Goldman Sachs
https://www.acast.com/exchangesatgoldmansac
hs/the-business-case-for-climate-action

Chapter 9

8. Naomi Klein (2014) *This Changes Everything:
Capitalism vs the Climate*; Allen Lane.

124. James Lovelock (2010) *A Final Warning: the
Vanishing Face of Gaia*; Penguin.

125. Robert Gifford (2015) "The Road to Climate
Hell". *New Scientist*, 11th July 2015.

126. Re-branded DECC, the Department for
Extreme Climate Change. Quoted in its
entirety from:
https://antidotecounteragent.wordpress.com/20
16/06/02/re-branded-decc-the-department-for-
extreme-climate-change/

ADDENDUM

Since I finished writing this book, there have been some major changes in the British political scene, some of which may impact on the issues I have discussed. First of all, the referendum about whether the UK should stay in the European Union has now taken place. The result was a marginal victory for those who wanted to leave and, despite much opposition internationally, from legal advisors and economists, from the 48.1% who voted to remain and from the millions of people who changed their minds and signed a petition asking for a second referendum, at the time of writing, it would appear that the UK will quit the EU.

Within three weeks of the referendum, the UK now has a new Prime Minister, a new Chancellor and the Cabinet is vastly changed. What effect will all of this have on the serious and urgent issue of climate change? Will the austerity measures of the last Chancellor, which had the effect of rewarding businesses to produce and sell abroad more goods, thus adding significantly to carbon emissions, be reversed? Will the EU's focus on renewable energy and the reduction of carbon emissions be abandoned by the UK altogether? Will the UK lose its European markets and seek new markets further afield, across the globe, thus adding further to carbon emissions associated with transport? Already there have been approaches from Australia, not surprising in view of the historical Australian antagonism towards Britain's presence in the EU, mentioned on page 74.

It is so difficult to tell what is going to happen. At the moment, every day there is surprising news.

What I would urge those who read this book, whether they agree with my politics or not, is to seriously lobby, both nationally and internationally for changes in the way we produce, sell and trade in goods and other economic products. Think twice before making decisions about their impact on the environment, on ecosystems and on the beautiful world that we enjoy today.

For tomorrow it may be lost.

Christine Parkinson

Lightning Source UK Ltd.
Milton Keynes UK
UKHW02f1336061217
313982UK00010B/196/P

9 781787 190412